Revisiting Plant Biostimulants

Edited by Vijay Singh Meena, Hanuman Prasad Parewa and Sunita Kumari Meena

Published in London, United Kingdom

Revisiting Plant Biostimulants
http://dx.doi.org/10.5772/intechopen.98125
Edited by Vijay Singh Meena, Hanuman Prasad Parewa and Sunita Kumari Meena

Contributors
Manoj Chaudhary, Abhijit Mandal, Soumyadarshi Muduli, A. Deepasree, Abshiba, Vaibhav Tripathi, Anil Kumar Sahu, Deepak Kumar Dash, Chandra Kishore Tyagi, Anjal Singh, Zenzile Peter Khetsha, Moosa Mahmood Sedibe, Rudolph Johannes Pretorius, Phoka Caiphus Rathebe, Karabelo Moloantoa, Mehmet Uğur Yildirim, Tuba Arjumend, Ercüment Osman Sarıhan

First published in London, United Kingdom, 2022 by IntechOpen
IntechOpen is the global imprint of INTECHOPEN LIMITED, registered in England and Wales, registration number: 11086078, 5 Princes Gate Court, London, SW7 2QJ, United Kingdom

British Library Cataloguing-in-Publication Data
A catalogue record for this book is available from the British Library

Additional hard and PDF copies can be obtained from orders@intechopen.com

Revisiting Plant Biostimulants
Edited by Vijay Singh Meena, Hanuman Prasad Parewa and Sunita Kumari Meena
p. cm.
Print ISBN 978-1-80355-552-2
Online ISBN 978-1-80355-553-9
eBook (PDF) ISBN 978-1-80355-554-6

Meet the editors

Vijay Singh Meena has worked on various aspects of soil aggregation, carbon management index, and carbon and nitrogen sequestration potential in the context of climate-resilient agriculture. He has identified the carbon management index as the key indicator for measuring soil degradation in different agroecosystems.

Hanuman Prasad Parewa specializes in soil fertility, INM, and plant growth-promoting rhizobacteria. His research has shown that the integrated application of fertilizer, FYM and bio-inoculants is highly effective for sustainable wheat and mung bean production and soil quality.

Sunita Kumari Meena has worked on the transformation and availability of phosphorus and sulfur in relation to soil organic carbon under long-term nutrient supply options. She is currently an assistant professor of soil science at the Sugarcane Research Institute (SRI), Dr. Rajendra Prasad Central Agricultural University, Pusa-Samastipur (Bihar), India. She has been a research scholar at both Banaras Hindu University, Varanasi, and at ICAR-Indian Agricultural Research Institute, New Delhi.

Contents

Preface

The agriculture sector is today facing the combined challenges of increasing productivity to feed the growing global population, and using resources efficiently while reducing the environmental impact on ecosystems and human health. Fertilizers and pesticides are powerful tools for growers to increase yields and guarantee continuous productivity. In recent decades, technological solutions to enhance the sustainability of agricultural production systems have been proposed, with a significant reduction in the use of synthetic agrochemicals like pesticides and fertilizers. A promising and environmentally friendly innovation would be the use of plant biostimulants (PBs) which enhance growth, development, crop productivity, and efficient nutrient use. The five chapters of this book explore a variety of eco-friendly technologies supporting sustainable production and address the impact of PBs on agricultural crops and medicinal and aromatic plants. The book will be useful to undergraduate and graduate students, teachers, and researchers, particularly in the fields of crop science, soil microbiology, and agronomy.

Vijay Singh Meena
CIMMYT – Borlaug Institute for South Asia (BISA),
Pusa, India

Hanuman Prasad Parewa
College of Agriculture,
Sumerpur, India

Agriculture University,
Jodhpur, India

Sunita Kumari Meena
Dr. Rajendra Prasad Central Agricultural University,
Pusa, India

Section 1

Impact on Agricultural Crops

Chapter 1

Agronomic Biofortification of Food Crops: A Sustainable Way to Boost Nutritional Security

Manoj Chaudhary, Abhijit Mandal, Soumyadarshi Muduli, A. Deepasree and Abshiba

Abstract

After the green revolution, there has been a substantial increase in the productivity of food crops. But the nutritional aspect of crops could not keep pace with the growing demand of the population. This has led to a rise in malnutrition problems, especially in developing countries, due to a lack of balanced nutrition. Agronomic biofortification, the process of increasing micronutrient content in food crops through agronomic approaches, is seen as an important process to improve the status of malnutrition in the world. It is seen as a quick, safe, and cost-friendly approach to provide iron, zinc, and other micronutrients in our everyday diet. Unlike molecular/genetic approaches, agronomic biofortification is done on existing crops and varieties and hence the product is easily accepted by the consumers. Approaches like integrated nutrient management (INM) based on soil test values, microbial application, foliar spray of nutrients, can substantially increase the level of micronutrients, vitamins, folic acid, etc. in our food. With sufficient research interventions and awareness programs, agronomic biofortification can serve as a tool to improve the nutritional status of the world.

Keywords: agronomic biofortification, malnutrition, micronutrients, foliar spray, INM

1. Introduction

Malnutrition, the devil of hidden hunger has already gained its ultimate importance after setting of Millennium Development Goals (MDGs) followed by Sustainable Development Goals (SDGs). The problem of malnutrition is reached in every corner of the Earth. Worldwide, it has been reported that around 2 billion people are affected by malnutrition [1]. Among them, nearly 850 million individuals experience the ill effects of undernourishment on this planet [2]. In low-income countries like Africa where the estimated risk for micronutrient deficiencies is high for Ca (54% of the continental population), Zn (40%), Se (28%), I (19%) and Fe (5%) [3]. Malnutrition mainly affects women and younger children in different forms in developing countries. An abysmal estimate of 151 million children under the age of 5 years are reported to be "stunted" and 51 million falls under the "wasting" category, that is, no proportionate weight as per the height [4]. 79.1% of India's children between the ages of

3 and 6 years, and 56.2% of married women (15–49 years) are anemic [5]. Vitamin A deficiency affects 169 million preschool children in South and Southeast Asia (33% of all preschool children) and 104 million (32% of all preschool children) in sub-Saharan Africa [6]. Various factors are responsible for malnutrition, but the unavailability of a balanced diet is the prime cause of it. The increasing deficiency of micronutrient in soil reduces the essential elements like minerals, vitamins in food and helps in malnutrition. Micronutrient deficiencies, even mild to moderate ones, can cause serious human health issues, such as impaired metabolic function, decreased immunity, and thus higher susceptibility to infections, growth failure, cognitive impairment, and, eventually, reduced productivity. Micronutrient deficiencies, even mild to moderate ones, can cause serious human health issues, such as impaired metabolic function, decreased immunity, and thus higher susceptibility to infections, growth failure, cognitive impairment, and, eventually, reduced productivity [1]. Hidden hunger can be prevented by direct (nutrition-specific) as well as indirect (nutrition-sensitive) interventions (**Figure 1**) [7].

Direct interventions focus on consumption behavior of humans and include dietary diversification, micronutrient supplementation, modification of food choices and fortification, whereas nutrition-sensitive interventions address the issue of malnutrition and include biofortification.

Fortification is a feasible, cost-effective, and sustainable practice for delivering the content of essential micronutrients, vitamins, and minerals (including trace elements) in the food, that improve the nutritional quality of the food and help to reduce the risk of public health problems. Biofortification, on the other hand, is the process of improving the nutritional quality of food crops using agronomic methods, traditional plant breeding, or modern biotechnology [8]. Biofortification differs from conventional fortification in that it tries to boost nutrient levels in crops during plant growth rather than using manual methods during crop processing. Biofortification may thus be a viable option for reaching populations where supplementation and traditional fortification methods are difficult to implement and/or limited [8]. Biofortification is primarily focused on staple crops which are starchy in nature like rice, wheat, maize, sorghum, millet, sweet potato, and legumes because they dominate diets worldwide, particularly among the groups which are vulnerable to micronutrient deficiencies, and

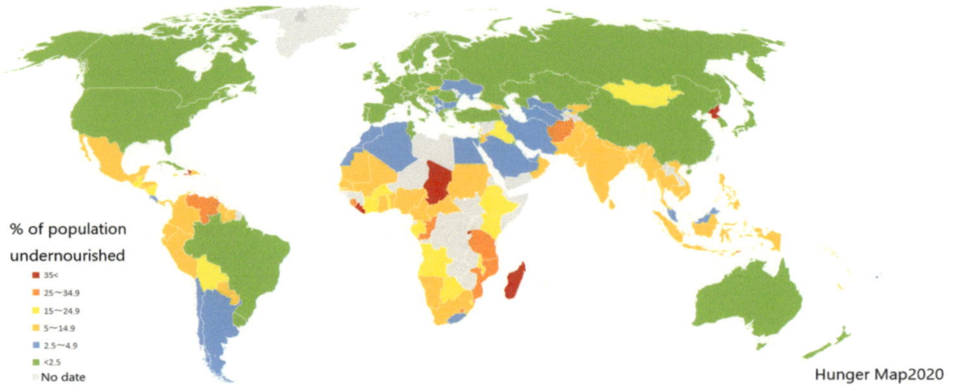

Figure 1.
Percentage of the population affected by undernutrition by country, according to United Nations statistics from 2012 (source: Wikipedia).

provides a feasible way of reaching malnourished populations with limited access to diverse diets, supplements, and commercially fortified foods [9].

The major drawbacks in biofortification through traditional plant breeding or genetic engineering is not only it require a long gestation period, adequate fund but also the products are not accepted in every country. Whereas agronomic biofortification is the easiest, fastest, and widely accepted way to reach the poorest of the poor rural masses and make foods rich in micronutrients, vitamins, Folic acids, etc. For example, Integrated application of AMF, P, and irrigation regimes on okra have given an increase in average total N, P, K, Ca, B and Mo uptake by 8, 24, 5, 14, 8 and 40%, respectively, over their non-AMF treatments [10].

2. Need for biofortification

Micro-nutrients are vitamins and minerals needed by our bodies in small quantities. However, their impact is critical, and its deficiencies create serious ill-health (WHO) like chronic diseases and stunting, weakening of immune system and reproductive systems and reducing our physical and mental abilities. More than 2 billion people suffer from micronutrient malnutrition and >20 million mortalities annually [11, 12]. It is also referred to as "hidden hunger". Among which Zn and Fe deficiencies rank 5th and 6th and mostly persisting in low-income countries (Ten leading causes of illness and disease in low-income countries, [13]. Children and women are most susceptible to micronutrient deficiencies. WHO estimates that, in 2017, over 6.3 million children under 15 years old and 5.4 million of them under 5, died as a result of malnutrition), particularly micronutrients [14]. This is mainly due to poor intake of proteins and micro-nutrients like Iodine, iron, Zinc, or monotonous food habit, lack of access to high-quality micro-nutrient-rich foods. Poor intake of micronutrient enriched food by pregnant mother' results in stunting of children when they were in the womb of the mother. Malnutrition is estimated to affect more than half of the world's population, making it one of humanity's most critical global concerns Conventionally industrial fortification and pharmaceutical supplementation are major steps for alleviating malnutrition issues. But these things are low reachability to poor income countries sometimes they reluctant to intakes of this tablet. So, the efficiencies of these strategies are low. So as an innovative step Biofortification introduced, it is an act of breeding nutrients into food crops, is a relatively low-cost, long-term way of increasing micronutrient delivery. This strategy not only reduces the number of severely malnourished persons who need complementary therapy but will also assist them in maintaining their improved nutritional condition. Moreover, Biofortification is a practical way to address impoverished rural people who may not have access to commercially available fortified foods and supplements. They have cereal-based food habit which has less protein and vitamin and soils of this region are low in Zn (50%), Fe (30%), and iodine, most of the soil is degraded due to alkalinity and salt issues [15]. Micro-nutrient deficiencies affect yield, the various metabolic functions of crops like seed formation, flowering, and quality of foodstuffs. Some micronutrients, especially B, Mg, and Cu are involved in cell wall stability and strength and thus increase plant resistance against pathogen penetration. So agronomic fortification is also a major concern of biofortification. fortification not only insists on intensifying micro-nutrient content but also increase the bioavailability of micronutrient and reduce the quantity of anti-nutritional factors.

Three main difficulties that must be addressed in order for biofortification to be successful:

i. A biofortified crop must be high yielding and profitable for the farmer;

ii. A biofortified crop must be efficacious and effective in reducing micronutrient malnutrition in humans; and

iii. A biofortified crop must be acceptable to both farmers and consumers in target regions [16].

3. Ways of biofortification

Because traditional therapies are ineffective, biofortification has been advocated as a long-term alternative for increasing mineral nutrition [17]. Biofortification is a method that improves both mineral content and bioavailability in the edible parts of staple crops. The former can be accomplished through agronomic intervention, plant breeding, or genetic engineering, whereas mineral bioavailability can only be influenced through plant breeding and genetic engineering (**Figure 2**).

4. Agronomic approaches for bio-fortification

A sufficient and balanced diet that supply the energy pathways and essential amino acids (lysine, methionine), vitamins (A, B, C, D and E), minerals, folic acids, ionic elements (Fe, Zn, I and Se) is possibly the most important contribution to human health and prophylaxis. Micronutrient deficiencies such as iron (Fe), zinc (Zn), iodine (I) and deficiency of vitamin A in soil and plants, which eventually appear as malnutrition in humans are one of the major causes of human disease

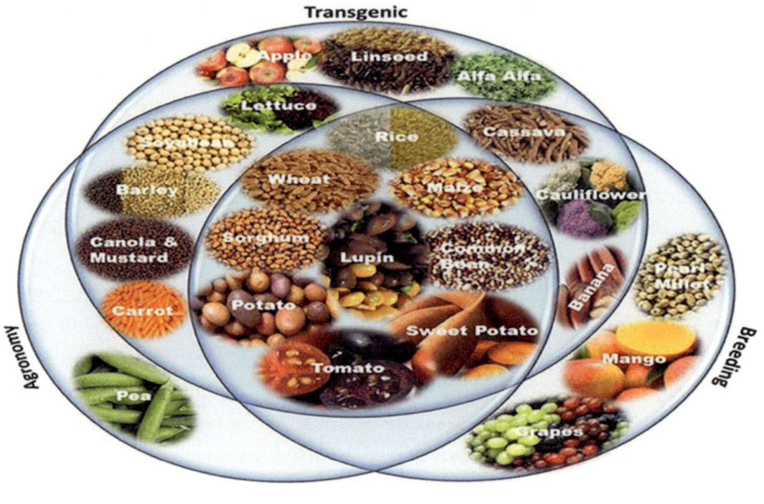

Figure 2.
Biofortified crops produced by different approaches-Breeding, Transgenic and Agronomic.

burden in the developing world. This results in severe impairments of human health and development and affects physical growth, immune system, cognitive development, maternal mortality, etc. A huge increase in food production must be achieved to feed the ever-increasing world population and to sustain human well-being. To meet the challenge of food security, agricultural production must be increased on the existing land, and therefore crop production must be intensified per unit of land.

There are 17 essential plant nutrients that are required by plants for their proper growth and development. Carbon (C), hydrogen (H), and oxygen (O) are not considered mineral nutrients but are the most abundant elements in plants and can be obtained from water and air. The remaining 14 are classified as macronutrients and micronutrients based on the relative requirement of these nutrients by plants. The macronutrients are nitrogen (N), phosphorus (P), potassium (K), sulfur (S), calcium (Ca), and magnesium (Mg). Compared with the macronutrients, the concentrations of the eight micronutrients iron (Fe), zinc (Zn), manganese (Mn), copper (Cu), boron (B), chloride (Cl), molybdenum (Mo), and nickel (Ni) are very small. Four additional elements sodium (Na), cobalt (Co), vanadium (V), and silicon (Si) have been established as beneficial micronutrients in some plants. If a single essential plant nutrient is available in insufficient quantity, it affects plant growth and thus the yield. Micronutrients are often referred to as minor elements, but this label does not mean that they are less important than macronutrients. Micronutrient deficiency or toxicity can reduce plant yield just like macronutrient deficiency or toxicity does, as they serve many important and critical functions in plant metabolism, growth and overall development of the plant.

Agronomic biofortification is the process of increasing the density of nutrients, vitamins and minerals in a crop by means of adopting proper agronomic practices and can be considered as an effective strategy for supplementation of micronutrients powders and enhancing dietary diversity.

The major advantages of agronomic biofortification are:

I. It is practiced on crop cultivars already being cultivated by the farmers and have good consumer acceptability of the produce

II. Enhanced micronutrient concentration in grain and other parts of the crop can be achieved in the same year

III. Very less amount of micronutrient is needed when the foliar application is followed

IV. No investment is needed for new seed

V. Agronomic biofortification always creates a win–win approach for developing countries.

The agronomic practices by which we can increase nutrient concentration in edible part:

1. Maintaining soil health physical, chemical, and biological properties

2. Proper cultivation practices

3. Balanced and integrated nutrient management

4. Other practices

4.1 Maintaining soil health physical, chemical, and biological properties

Soil health is one of the important factors regulating plant health by providing optimum conditions like proper root growth, increasing availability of nutrients, moisture-holding capacity and biological activity, optimum aeration, etc. for plants to grow which helps to increase nutrients concentration on edible parts.

4.1.1 Soil physical properties

The availability of micronutrients significantly depends on soil texture. Sandy soil has fewer micronutrients compared to fine-textured soils, because of its high water and nutrient holding capacity. In case of reduced moisture condition rates of dissolution and diffusion of nutrients get reduced and root activity is also reduced. Good soil structures like loose, crumby, and granular possess good porosity, less compaction, high nutrient, and moisture-holding capacity, increase nutrient uptake compared to platy and blocky structured soils, which helps proper root growth hence providing more qualitative yield. Application of organic matter, amendments, press mud, tank silt, bentonite clay can improve soil structure, as well as they increase water and nutrient holding capacity which helps to increase nutrients, folic acid, vitamins in crops.

Though submergence can increase Fe, Mn increases but the uptake of other macro and micronutrients are reduced. So proper drainage facilities are also essential for nutrient uptake and translocation to grains, bulbs, stems, fruits, etc.

4.1.2 Soil chemical properties

Soil chemical properties also possess their impact on maintaining quality and quantity crop products. If the soil has a high buffering capacity, it can provide or resupply more nutrients to the crops. Soil CEC and AEC also possess a great impact on both micro and macronutrient availability. Soils with high CEC hold more nutrients and provide them when the crop needs them. Increased base saturation in soils increases the availability of nutrients like Ca, Mg, K and other cations. Nutrients also become less available to crops in too acidic or too alkaline conditions. In acidic conditions availability of calcium magnesium potassium declined but in alkaline conditions Mn, Zn, Cu may become less available. Availability of Phosphorus is less at too low or too high pH.

Proper soil chemical and physical properties are needed for successful biofortification. Application of gypsum, sulphur compounds on alkaline soil, and lime on acid soils can help to maintain soil chemical properties.

4.1.3 Soil biological properties

Optimum biological activity is needed for faster mineralization of the nutrients. An increasing number of microorganisms like PGPR, AMF, and mycorrhiza act as an extension of the root system and can mobilize or solubilize both mobile and immobile nutrients and make them available to plants. Some macro-organisms like an earthworm, mole cricket, ants make the soil more porous and help the roots to penetrate deeply. So biological activity plays a crucial role in biofortification in a sustainable way. Application of organic matter, the addition of legume crops in the cropping system, less use of pesticides can increase biological activity in the soil.

4.2 Proper cultivation practices

4.2.1 Tillage

Tillage is an important practice for most crops. Proper tillage can provide the most suitable soil conditions where crops can germinate, grow up and complete the life cycle. Tillage eliminates weeds, disease inoculants and provides a competitive advantage to the crops. Tillage at optimum moisture conditions (i.e., 50–75% MHC) is crucial for tillage operation as more or less moisture can create hardpan in subsoil which restricts root growth and hence reduce nutrient uptake and yield. Now a days reduced tillage, or zero tillage is gaining its importance, but soil compaction is the main problem for them as it creates problems in root proliferation [18]. Stipesevic et al. [19] reported that in winter wheat Zn concentration in the plant tissue at the beginning of heading did not differ due to tillage treatments in the first 2 years, but in the third year it was 11.7 mg kg^{-1} in the conventional tillage plots and only 6.4 mg kg^{-1} in the zero-till plots. Subsoil or Chisel plow once in 3–4 years is a solution for them. Some improved tillage practices like a ridge and furrow planting, Furrow irrigated raised bed planting (FIRB) also help to increase the nutrient uptake by the crops.

4.2.2 Water management

As most of the nutrient uptake is done by mass flow and diffusion so soil moisture is the main factor that affects nutrient concentration in crop products. Optimum moisture helps in better root growth, increases the solubility of nutrients, and makes it available to the plants. Both excess and deficit water reduce nutrient concentration from the root zone by leaching or restricting mobilization. Sometimes mild stress can increase nutrient concentration in grains. Water deficit during grain filling can decrease lipid content in wheat grains but mild water deficit would be beneficial to the grain filling and starch compositions, significantly improving bread-making quality [20]. Proper management of water in the wheat field at the post-harvest stage was helpful both to improve protein content and composition of wheat grain, but water deficit/water stress at the pre-anthesis stage can increase P, Ca, Mg, K, and Zn. Proper management of water in all the critical stages is important for improving the quality of the product. Continuous flooding throughout the rice-growing season reduces Cu and Zn plant availability while increasing B, Fe, and Mn availability in both limed and un-limed acid laterite and alluvial soils. In comparison to continuous flooding under above-ground soil conditions, alternate flooding and drying were shown to be favorable to rice because it considerably enhances the availability of B, Cu, and Zn nutrients to plants while decreasing the availability of Fe and Mn nutrients.

4.3 Balanced and integrated nutrient management

Nutrient application is the most important step for the agronomic ways of biofortification. Integrated use of compost, manure, organic and inorganic fertilizer, microorganisms is the best way for a sustainable way of biofortification. Here we will discuss these things.

4.3.1 Application of organic matter

Soil organic matter influences greatly soil physical, chemical, and biological properties. It improves soil structure, soil porosity, bulk density, helps in stabilizing soil aggregates and other soil physical properties. For alkaline and saline soils, it also acts as a reclaiming agent. Besides improving soil health, it also has the capacity to supply all other nutrients to plants. Fe which is largely present in the insoluble form as Fe^{3+}, organic matter can increase its solubility through the effect of redox potential [21]. Fulvic acids, humic acids which are formed during the decomposition of organic matter help to increase Fe solubility and its availability to plants. Whereas other nutrients like Cu, Ni are tightly bonded with the organic fraction of soil which makes them less available. The addition of green manure, compost, biosolids, and biochar-causes more uptake of soil-bound Zn and other nutrients and intensifies the plant availability of zinc [22]. These amendments also decrease heavy metal uptakes like Cd in rice [23]. The addition of organic matter shows a considerable increase in microbial biomass carbon, microbial community diversity. These biological properties of soils may help to maintain nutrient cycling and soil quality. The foods grown in organic conditions have greater nutrient content including minerals and vitamins [24]. So, we can blindly say that if we want successful biofortified crop products by the agronomic way, organic matter is the only solution.

4.3.2 Application of synthetic fertilizers

Application of macronutrients like N, P, and K is recommended based on soil test values and nitrogen should be applied in split doses. These nutrients promote root and shoot growth and increase uptake of all nutrients by the plants. Intensive use of macronutrient fertilizers sometimes supplies micronutrients as micronutrients are added with these during manufacturing process or present as impurities. High doses of nutrients like N, P, and K reduce the uptake of nutrients which has low phloem mobility like Ca, as Ca is prone to dilution effects [25]. Over-reliance on ammonium-based fertilizers limits cation nutrient uptake and decreases the carbohydrate content of root vegetables by increasing root respiration [26]. Excess soil P causes more phytate content and can promote Zn deficiencies. Whereas excess consumption of K intervenes Ca and Mg uptake [26]. So, judicious use of macronutrients is most important to help in the proper uptake of other nutrients.

4.3.3 Micronutrient application and bioavailability

Micronutrients simply follow a straight pathway to reach into the human body from the soil through the crop and food. The success of agronomic biofortification in alleviating micronutrient deficits in humans is determined by several important parameters. The parameters are mostly influenced by nutritional bioavailability at various stages (**Figure 3**).

Soil application of micro-nutrients can increase grain nutrient content like soil application of Zn increase Zn concentration in cereal crops for 2–3 times depending on crop species [27] and crop genotype [28]. In basmati rice grain and straw, green manure and Zn-coated fertilizers enhanced nutritional content and absorption. Foliar fertilization of 0.2 % zinc sulfate recorded a higher Zn concentration in rice, whereas Zn-coated urea (ZCU) as $ZnSO_4.H_2O$ registered the highest total Zn uptake [29].

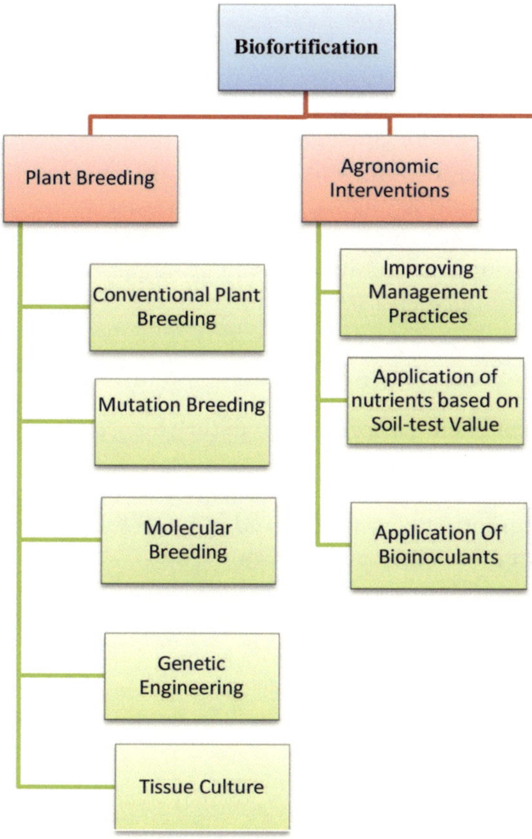

Figure 3.
Ways of biofortification.

Kaur [30] found a considerable increase in micronutrient uptake (Zn, Fe, Mn, Cu, and B) in wheat after applying 100 percent P, 10 kg Zn, and 1 kg B ha^{-1}. Kumar et al. [31] reported that increasing the application of boron levels from 0.5 to 1.5 kg ha^{-1} should reduce B use efficiency and the highest value (9.2%) was obtained at a lower level of applied B (0.5 kg ha^{-1}), whereas the lowest was found (4.2%) with B applied at 1.5 kg ha^{-1}. The foliar spray helps to transport the nutrients from the site of application to the site of utilization in a rapid way. Fe, Zn, and Mn are applied in chelated form and translocation within the plants was found greater [32]. Foliar fertilization with $ZnSO_4$ and Zn EDTA and other chelates has been used in fruits and vegetable production. From these vegetative parts, nutrients will translocate to the edible parts.

In rice, Zn and Fe are localized in protein bodies in the outer layer of the grains, which is often removed during processing (de-husking, milling) leaving less Zn and Fe in the consumed rice [33, 34]. Rice parboiling is an effective method to increase nutrient contents especially when micronutrients are added to the soak water during the parboiling, as the process drives nutrients from the bran and germ layer to the endosperm [35, 36].

Application of 120 kg Si ha^{-1} increased rice yield to the tune of 17.1%, 7.1% and 2.0%, respectively, over 0, 40 and 80 kg Si ha^{-1} [37]. So, we can say that only application of micronutrients is not sufficient for successful biofortification, its bioavailability also needs to take into consideration.

4.3.4 Through the application of Microorganisms

The most active site for the soil microorganisms in the rhizosphere where the nutrients are sequestrated, mobilized, and made available to plants. Bio-fortification of the crops can be done by using bio-fertilizer or microbial inoculants which mobilize or solubilize the essential nutrients and possess a positive impact on plants' health.

Most organisms possess both direct and indirect effects for improving plant health and nutrient concentration on grain and biomass. The microorganisms like PGPR, AMF fungi, Cyanobacteria, Actinomycetes are the major drivers.

4.3.4.1 Role of PGPR

PGPR helps to increase the nutrient concentration in the rhizosphere in different ways:

I. They release growth-promoting compounds and mineral solubilizing enzymes which play an important role in the cycling of nutrients

II. They modify root morphology and thereby increase the root surface area which helps in more nutrient uptake

III. Sometimes they secret Phyto-siderophores which increases micronutrient availability in soil

Inoculation of bacterial strains like *Pseudomonas putida, Pseudomonas fluorescens, Azospirillum lipoferum* increase iron concentration up to two to three times in rice [38]. Rana et al. [39] observed that the treatment involves *Providencia* sp. bacteria can increase zinc copper Ion concentration in wheat grain. Santiago et al. [40] found that Fe concentration in the biomass of wheat could increase up to 1.5-fold when the plot is treated with a siderophore-producing strain *Trichoderma asperellum*. Tariq et al. [41] reported that commercial application of *Pseudomonas sp.* in rice soil improve Zn concentration up to 157% in rice. *Pseudomonas sp.* and *Actinobacteria* sp. inoculation improve uptake of Fe, Mg, Ca, K and P by crop plants [42].

4.3.4.2 Role of Fungi

Most of the fungi are being heterotroph (saprotrophs, biotrophs and necrotrophs) in nature so they play an important role in regulating soil fertility by decomposing and cycling of organic matter and minerals.

Arbuscular mycorrhiza has an extensive hyphal network that spread both internally and externally in the roots. They explore the soil more efficiently as their hyphae have some specific characteristics like faster growth rate, thin and extensive branches. AM fungi can increase forage area up to 100 times as compared the root length of the crop. AMS has the ability to improve the supply of N, P, Cu, Zn, Fe, Ca, B, Mn, Ni, K, etc. [43].

Some Ecto-mycorrhizal fungi also produce low molecular weight organic acids that help more nutrient mobilization.

Due to the application of AMF+ P+ Proper irrigation in okra total N, P, K, Ca, B and Mo uptake was increased 8, 24, 5, 14, 8 and 40%, respectively whereas in the case of pea, an increasing amount of total N (8%), P (19%), K (12%), Mg (12%), Ca (22%), Zn (22%), Fe (10%), Cu (28%), Mn (10%), B (11%) and Mo (38%) uptake was also addressed in AMF imbedded treatments over non-AMF counterparts [10].

4.3.4.3 Role of Cyanobacteria

Cyanobacteria or blue green algae is the plant growth-promoting agent which is also a major player in nutrient uptake and improving user efficiency. They increase nutrient concentration in plants by:

I. The counter deleterious pathogenic activity and maintain good plant health.

II. They produce allelochemicals like IAA, extracellular polysaccharides which stabilize the soil and increase N and C in the rhizosphere regions.

III. They help in sequestering nutrients and improving their mobilization into plants.

When Anabaena-based biofilm inoculants were used in rice soils under flooded and SRI methods of rice cultivation that increases 13–46% iron and 15–41% zinc in rice grains respectively. In Anabaena-Pseudomonas-based biofilm treatments, rice grains showed an increase in copper accumulation.

Cyanobacterial inoculation helps to increase rice crop yields (grain yields) to the extent of 10–24% in diverse locations in the world, especially in South Asia [44].

4.3.4.4 Role of Actinomycetes

Actinomycetes can play a significant role to dissolve the primary rock-forming minerals to obtain essential nutrients and act as nucleation sites for the precipitation of secondary minerals. In this way, it helps to uptake nutrients by plants (**Figure 4**)

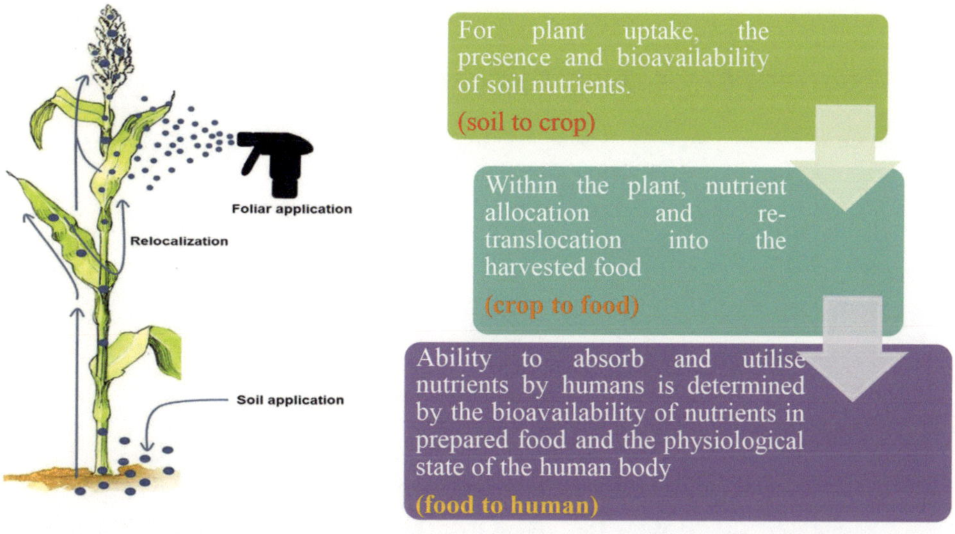

Figure 4.
Agronomic biofortification is the application of micronutrient-containing mineral fertilizer (blue circles) to the soil and/or plant leaves (foliar), to increase micronutrient contents of the edible part of food crops.

Pulses	Energy (kcal/100 g)	Vitamin (mg/100 g)				Nutrient (mg/100 g dry weight)				
		Thiamine (mg)	Riboflavin (mg)	Niacin (mg)	Vitamin C (mg)	Fe	Zn	Ca	Mg	K
Chickpea (*Cicer arietinum L.*)	368	0.5	0.2	1.5	4	6.2	3.4	105	115	875
Pigeon pea (*Cajanus cajan L.*)	342	0.6	0.18	2.9	—	5.2	2.7	130	183	1392
Lentil (*Lens culinaris*)	346	0.8	0.2	2.6	4.4	7.5	4.7	56	122	955
Mung bean (*Vigna radiata L.*)	345	0.6	0.2	2.2	4.8	6.7	2.7	110	189	1246
Urdbean (*Vigna mungo L.*)	347	0.6	0.2	2.3	4.8	8.4	3.5	55	—	—
Field pea (*Pisum sativum L.*)	345	0.7	0.2	2.9	—	4.4	3.0	186	115	981
Rajmash (*Phaseolus vulgaris L.*)	345	0.53	0.22	2.08	1.8	3.4	1.9	80.3	188	1316
Cowpea (*Vigna unguiculata*)	346	0.94	0.22	2.36	4.6	7.54	3.77	287	250	1450
Horse gram (*Macrotyloma uniflorum*)	321	0.4	0.2	1.5	—	7.0	—	202	—	—
Moth bean (*Vigna aconitifolia*)	330	0.4	0.09	1.5	—	9.6	—	—	—	—

Table 1.
Nutritional profile of pulse grains.

5. Other practices

5.1 Crop rotation

The beneficial effects of crop rotation include improved soil chemical and physical fertility, reduced weed infestation and diseases. Karlen et al. [45] concluded that crop rotation and cover crops may increase the availability of Fe, Cu and Zn.

In rice-wheat rotation use of FYM and green manurer maintained the available fraction of soil micronutrients like Fe, Zn, Cu and Mn compared to the same rotation fertilized with inorganic fertilizer alone [46].

The addition of pulse crops in the cropping system is the best option after cereals for improving eating quality, not only because of their importance for humans and animals but also due to their soil ameliorative values and their ability to thrive under harsh and fragile environments (**Table 1**).

5.2 Intercropping

Intercropping between soil exhaustive crop and the regenerative crop can create a complementary relationship and helps to reduce weed and disease infestation, protect the soil from nutrient mining, maintain soil physical, biological health and helps to increase nutrient density on them.

5.3 Proper pest management

Pests like insects, weeds, disease inoculants possess a great impact on the quality as well as quantity of the product. They restrict the growth of the crops, sometimes can kill the plants. They also create a bitter taste in plants by producing some toxins. So proper management of them is of utmost importance. Integrated pest management is the best option to control their infestation as well as to maintain the quality of the product.

5.4 Proper drying and storage

During post-harvest season grains that are not properly dried can sometimes develop mold and also some toxic substances like aflatoxins, ochratoxins, so proper

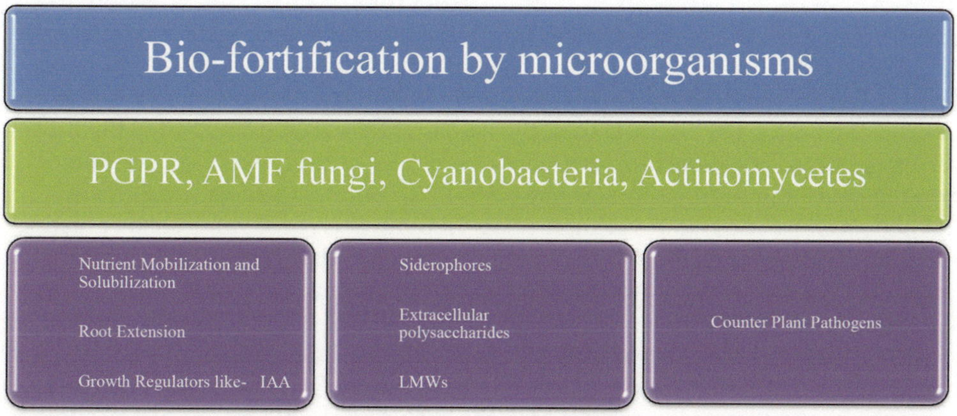

Figure 5.
Overview of mechanisms involved in microbe-mediated biofortification.

Crops		Type of biofortification done through agronomic approaches
Cereals	Rice	Fe, Zn, Se
	Wheat	Fe, Zn, Se,
	Maize	Fe, Se, Zn
	Barley	Zn, Fe
	Sorghum	Protein
Pulses	Soybean	Se
	Chickpea	Zn, Se, Fe, Zn, Ca, Cu, Mn, Mg
	Pea	Zn
	Common bean	Zn N, P, K, Cu, Mn, Zn
Oilseeds	Canola	Protein, oleic acid, linoleic acid
	Mustard	Se
Vegetables	Potato	Zn, Se
	Sweet Potato	Beta-carotene
	Carrot	Iodine, Se
	Lettuce	Iodine, Se
Fruits	Tomato	Iodine

Table 2.
Type of biofortification done on crops through agronomic approaches.

drying is necessary. The grains like rice and wheat are exposed to contaminants, pests and diseases and prone to nutrient losses. So proper storage is important after harvest (**Figure** 5 and **Table** 2).

6. Progress on biofortification

As of 2018, worldwide 6.7 million farm households are producing biofortified crops and these products surely go into food dishes. Till now more than 300 varieties have been released in 30 countries for crops, such as rice, wheat, maize, cassava, orange sweet potato, potato, lentil, beans, cowpea, banana, and plantain [47]. Several institutions like 1. Food Policy Research Institute (IFPRI), Bill and Melinda Gates Foundation (BMG Foundation), Biotechnology Industry Research Assistance Council (BIRAC), Indian Agricultural Research Institute (IARI) must work together to populate biofortified crops and create an enabling environment. Recognition of biofortification among global regulatory agencies, a collaboration between agencies from various sectors, a more active role for private players, and designing new development policies and agendas that take into account the programs currently being implemented on the ground, among other things, are all components of such an environment. CGIAR will continue to employ its varied network of international organizations, research institutes, and civil society organizations around the world to drive a single, integrated conversation on standards and governance, and to provide society with the highest possible return on investment. Harvest Plus is one of them, and it is leading the biofortification project, which it will enable in the next years, with local governments acting as main partners [47].

7. Constraints in agronomic biofortification

Enhancement of crop qualities through agronomic biofortification has the following challenges:

- Timely availability to farmers—Lack of availability of micronutrient fertilizers at the proper time to the farmers leads to farmers mostly skipping their application to the crops, which further leads to widespread deficiencies.
- Low nutrient use efficiency for micronutrients—Micronutrients like iron, zinc, copper, etc. have very low use efficiencies (1–5%) which limit the uptake of applied micronutrients by plants.
- Genetic constraints—Agronomic biofortification has a minor role in enhancing protein content as both are negatively correlated and protein content is genetically controlled.
- Difficulty in public awareness—Iron and zinc deficiencies are widespread in India and around the world. Since their deficiencies stay hidden and are not easily manifested as external symptoms, the creation of public awareness about the adverse effects of iron and zinc malnutrition is important.
- Lack of knowledge—In most crops, a thorough understanding of the mechanisms of mineral translocation from soil to plant is inadequate. As a result, further information regarding the rate-limiting processes of micronutrient acquisition and translocation in the soil-plant system is required.
- Safety in the use of biofortified crops—The safety concerns of biofortified crops have to be analyzed in detail before making them available in the market. A comprehensive knowledge gap also exists in the bioavailability of micronutrients in food grain and mineral distribution patterns in plant systems.
- Post-harvest processing losses—The loss of micronutrients after harvesting, on processes like selective removal of outer tissues during cleaning and processing is not analyzed for most of the crops and needs to be considered.

8. Future prospects

The public sector institutions must give intensive efforts and make policy for promotional campaigns that can significantly increase the acceptance of agronomic practices for biofortification. Providing the micronutrient fertilizers and other bio-inoculants like PGPR, AMF, cyanobacteria can cause the rapid spreading of these agronomic practices. Assured premium remunerative prices for the biofortified products in the market encourage farmers to grow more biofortified foods. Active investment of extension activities would create awareness among farmers' industries and consumers regarding the availability and benefits of these biofortified crops (**Figure 6**).

Some essential steps should be required for the popularization of biofortified crops. These are:

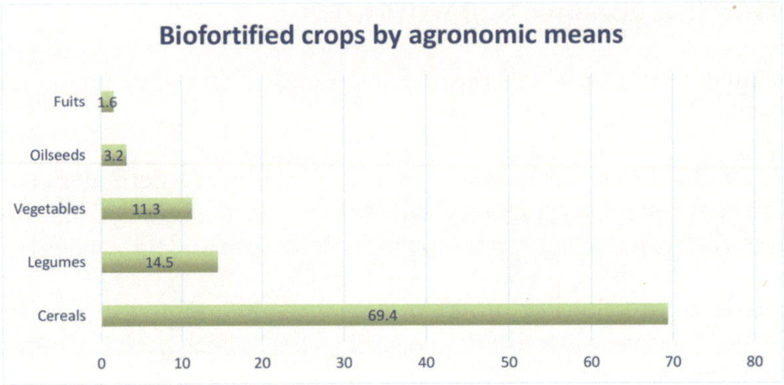

Figure 6.
Percentage of biofortified crops by agronomic means [48].

8.1 Awareness generation

Incomplete knowledge of the health benefits of biofortified crops is the major reason for slow adoption. Some demonstration trials should be conducted in the farmer's field and make sure that farmers would actively participate in this program. Lack of yield compared to conventional practices is also another reason for slow adoption but if farmers get premium prices, they do not lose their interest in this. Strong linkages should be constructed with Agro-processing industries that provide confidence to farmers. Strong promotional extension activities such as field demonstration, conveying a message through TV talk, radio shows and live drama would make the farmers, industry, and consumers aware of the existence and benefits of biofortified crops.

8.2 Policy support

Strengthening input supply is a major step towards the popularization of Biofortified crops. Providing subsidized micronutrient fertilizer, bio-inoculants, or microorganisms, receiving to provide remunerative prices for biofortified grains in the market will encourage farmers. Recently, unveiled National Nutrition Strategy—2017 by NITI Aayog, the Government of India envisages the alleviation of malnutrition in the country through food-based solutions [49].

Inclusion of this biofortified cereal indifferent government-sponsored programs such as National Food Security Mission, Rashtriya Krishi Vikas Yojna as well as nutrition intervention program such as Integrated Child Development Services scheme, 'Mid-day meal' and Nutrition Education and Training through Community Food and Nutrition Extension Units would help in providing the much-needed balanced food to poor people. Recently, the Government of India announced the millets like (sorghum, pearl millet, foxtail millet, finger millet, Kodo millet, proso millet, little millet, and barnyard millet), besides two pseudo millets (buck-wheat and amaranthus) as 'Nutri Cereals' which have high nutritive values. This would increase their demand in both the regional and Challenges to reach billion people by 2030 worldwide markets, allowing farmers to command better prices. Incorporating biofortified items into these government-sponsored programs would assist youngsters, pregnant women, and the elderly, as well as speed up their distribution. Given the well-documented health benefits of QPM, Ethiopia's government has set a goal of cultivating QPM varieties on 20% of the country's total maize land in the future years [50]. As a result, significant government policy support would improve the uptake and acceptance of biofortified crops (**Figures** 7 and **8**).

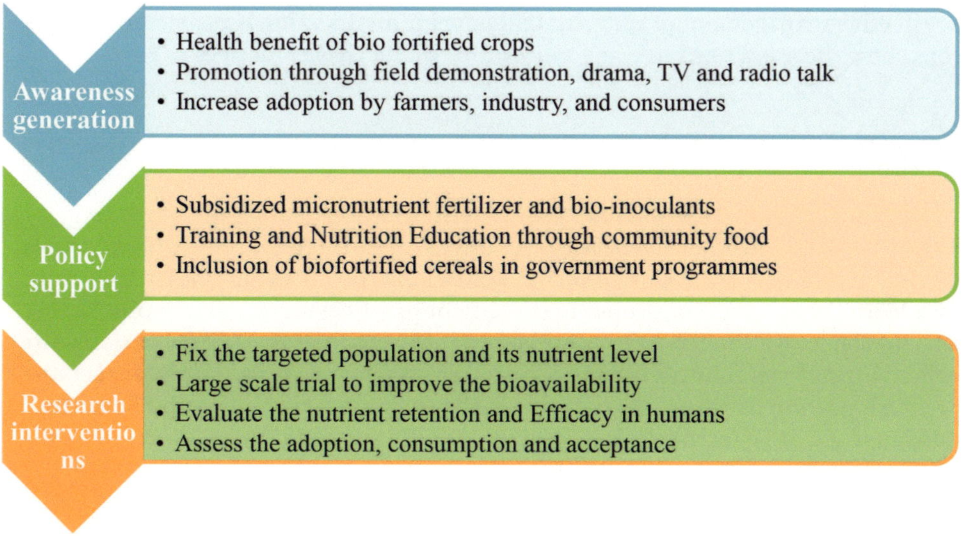

Figure 7.
Conceptual diagram on future prospects of agronomic biofortification.

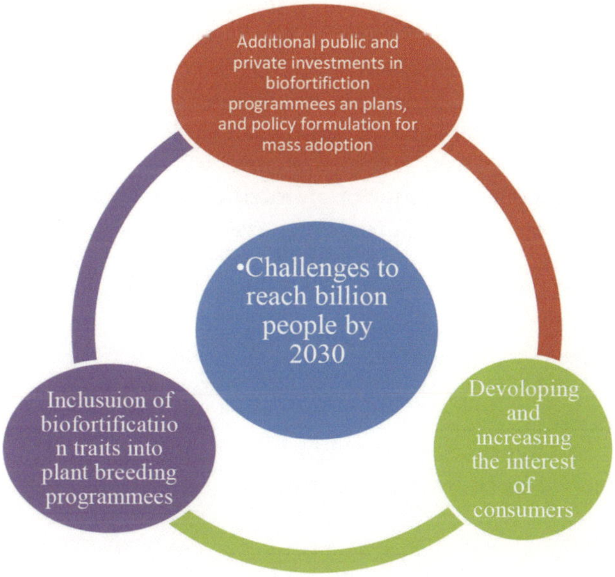

Figure 8.
Ways to reach biofortified products to people [50].

8.3 Research interventions

Most nutritional characteristics like (protein, lysine, tryptophan, iron, zinc, and Vitamin c) are invisible. So, it is difficult to convince farmers and Consumers regarding the quality of the product. Large-scale trials should be needed by a public institution that could help in generating more data and thus help in the dissemination of fortified crops research should be needed on nano-fertilizers that have the potential for effective management of nutrients till now biofortification is restricted to some crops show there is a need to bring more crops under these practices.

9. Conclusions

Agronomic approaches provide a short-term solution compared to breeding approaches. The introduction of high-yielding varieties and extreme use of commercial single fertilizers are the main reason behind micronutrient malnutrition problems. With the adoption of proper management practices significant improvement in nutrient concentration has been observed by different scientists. Fertilization with both micro and macronutrients has been reported to increase the nutritional status of the edible portion of a crop. An increase in the concentrations of Zn and Fe with the addition of Zn or Fe fertilizers has been reported in wheat (*Triticum aestivum*) [26]. Foliar application of Fe and Zn fertilizers has been found to be an easy and effective way of yield and nutrition enhancement in fruits and vegetable crops besides cereals [51]. Water management in winter wheat at the post-anthesis stage was helpful for improving grain quality and nutrient content relevant to the processing and human consumption [20] and the addition of organic matter in the form of green manure, compost, biosolids and biochar caused more uptake of soil-bound Zn and other nutrients and intensify the plant availability of zinc [21] has also been reported. An increase of Fe concentration up to 1.5-fold in the wheat biomass has been found by Watson et al. [22]. Santiago et al. [40] when the plot is treated with a siderophore producing strain T*richoderma asperellum*. To feed the ever-increasing population from the limited land resources require proper knowledge of soil-plant interactions and precise information on the status of different agro-ecological regions so that people can get quality food in their dish. In the short term, agronomic approaches are the most important sustainable technique of biofortification. Biofortification has expensive time-consuming regulatory approval processes, and its acceptance is very low in the society. Besides these challenges, biofortified crops hold a very bright future as these have the potential to remove micronutrient malnutrition among billions of poor people, especially in developing countries.

Conflict of interest

The authors declare no conflict of interest.

Author details

Manoj Chaudhary*, Abhijit Mandal, Soumyadarshi Muduli, A. Deepasree and Abshiba
ICAR-Indian Agricultural Research Institute, Gauria Karma, Hazaribag, Jharkhand, India

*Address all correspondence to: manoj310975@gmail.com

References

[1] Tulchinsky T. Micronutrient deficiency conditions: Global health issues. Public Health Reviews. 2010;**32**:243-255

[2] United Nations Millennium Development Goals Report. 2006. Available from: https://www.un.org/zh/millenniumgoals/pdf/MDGReport2006

[3] Joy EJM, Ander EL, Young SD, Black CR, Watts MJ, Chilimba ADC, et al. Dietary mineral supplies in Africa. Physiologia Plantarum. 2014;**151**:208-229

[4] Ramadas S, Vellaichamy S, Ramasundaram P, Kumar A, Singh S. Biofortification for Enhancing Nutritional Outcomes and Policy Imperatives. Elsevier; 2020. pp. 310-312

[5] Krishnaswamy K. The problem and consequences of the double burden—A brief overview. Programme and abstracts. In: Symposium on Nutritional Security for India-Issues and Way Forward. New Delhi: Indian National Science Academy; 2009. pp. 5-7

[6] IRRI. Bringing Hope, Improving Lives: Strategic Plan 2007-2015. Vol. 61. Manila, Philippines: International Rice Research Institute; 2006

[7] Ruel MT, Alderman H. Nutrition-sensitive interventions and programmes: How can they help to accelerate progress in improving maternal and child nutrition? And the maternal and child nutrition study group. Lancet. 2013;**382**:536-551

[8] World Health Organization. Biofortification of Staple Crops. 2019. Available from: https://www.who.int/elena/titles/biofortification/en/

[9] Saltzman A, Birol E, Bouis H, Boy E, De Moura F, Islam Y, et al. Biofortification: Progress toward a more nourishing future. Global Food Security. 2013;**2**:9-17

[10] Kumar A. Phosphorus and rain-harvested water economy through vesicular arbuscular mycorrhizae (VAM) in okra–pea sequence [PhD thesis]. Palampur: CSK HPKV; 2012

[11] Kennedy GL, Pedro MR, Seghieri C, Nantel G, Brouwer I. Dietary diversity score is a useful indicator of micronutrient intake in non-breast-feeding Filipino children. The Journal of Nutrition. 2007;**137**:472-477

[12] WHO. WHO/FAO (World Health Organization/Food and Agriculture Organization of the United Nations). Preparation and Use of Food-Based Dietary Guidelines. Nutrition Programme. Geneva: World Health Organization; 2003

[13] World Health Organization. The World Health Report: Reducing Risks, Promoting Healthy Life. 2002

[14] WHO Newsroom. Factsheet “Children: Reducing Mortality”. 2019. Available from: https://www.who.int/newsroom/fact-sheets/detail/children-reducing-mortality

[15] Fuge R, Johnson CC. Iodine and human health, the role of environmental geochemistry and diet, a review. Applied Geochemistry. 2015;**63**:282-302

[16] Hotz C, McCafeerty B, Hawkes C, Ruel M, Babu S. From harvest to health: Challenges for developing biofortified staple foods and determining their impact on micronutrient status. Food and Nutrition Bulletin. 2007;**28**:271-279

[17] Zhu C, Naqvi S, Gomez-Galera S, Pelacho AM, Capell T, Christou P. Transgenic strategies for the nutritional enhancement of plants. Trends in Plant Science. 2007;**2007**(12):548-555

[18] Busscher WJ, Sojka RE. Enhancement of sub-soiling effect on soil strength by conservation tillage. Transactions of ASAE. 1987;**30**:888-892

[19] Stipesevic B, Jug I, Tolimir M, Cvijovic M. Winter wheat and soybean zinc uptake in different soil tillage system. Cereal Research Communications. 2009;**37**:305-310

[20] Zhao FJ, Su YH, Dunham SJ, Rakszegi M, Bedo Z, McGarth SP, et al. Variation in mineral micronutrient concentration in grains of wheat lines of diverse origin. Journal of Cereal Science. 2009;**49**:290-295

[21] Lindsay WL. Iron oxide solubilization by organic matter and its effect on iron availability. Plant and Soil. 1991;**130**:27-34

[22] Watson CA, Oborn OI, Edwards AC, Topp CFE, Walker RL. Using soil and plant properties and farm management practices to micronutrient composition of food and feed. Journal of Geochemical Exploration. 2012;**121**:15-24

[23] Liu L, Chen H, Cai P, Liang W, Huang Q. Immobilization and phytotoxicity of Cd in contaminated soil amended with chicken manure compost. Journal of Hazardous Materials. 2009;**163**:563-567

[24] Woese K, Lange D, Boess C, Bogl KW. A comparison of organically and conventionally grown foods—Results of a review of the relevant literature. Journal of the Science of Food and Agriculture. 1997;**74**:281-293

[25] Marcelle R. Mineral nutrition and fruit quality. Acta Horticulture. 1995;**383**:219-226

[26] Marschner H. Mineral Nutrition of Higher Plants. 2nd ed. London: Academic Press; 1995

[27] Rashid A, Fox RL. Evaluating internal zinc requirements of grain crops by seed analysis. Agronomy Journal. 1992;**84**:469-474

[28] Graham RD, Ascher JS, Hynes SC. Selecting zinc efficient cereal genotypes for soils of low zinc status. Plant and Soil. 1992;**146**:241-250

[29] Pooniya V, Shivay YS. Enrichment of basmati rice grain and straw with zinc and nitrogen through fertifortification and summer green manuring crops under Indo-Gangetic Plains of India. Journal of Plant Nutrition. 2013;**36**(1):91-117

[30] Kaur R. Phosphorus and rain-harvested water economy through vesicular arbuscular mycorrhizae (VAM) in okra–pea sequence [MSc thesis]. Palampur: CSK HPKV; 2012

[31] Kumar A, Parmar DK, Suri V. Effect of boron fertilizers and organic manure on autumn cauliflower in western himalayas. Annals Horticulture. 2012;**5**(1):17-24

[32] Ferrandon M, Chamel AR. Cuticular retention foliar absorption and translocation of iron, manganese and zinc supplied in organic and inorganic form. Journal of Plant Nutrition. 1988;**11**:247-264

[33] Haas JD, Beard JL, Murray-Kolb LE, Del Mundo A, Felix A, Gregorio G. Iron biofortified rice improves the iron stores of nonanemic filipino women. The Journal of Nutrition. 2005;**135**:2823-2830

[34] Zimmermann MB, Wegmueller R, Zeder C, Chaouki N, Biebinger R, Hurrell RF, et al. Triple fortification of salt with microcapsules of iodine, iron, and vitamin A. The American Journal of Clinical Nutrition. 2004;**80**(5):1283-1290

[35] Prakash O, Ward R, Adhikari B, Mawson J, Adhikari R, Wess T, et al. Synchrotron X-ray fluorescence microscopy study of the diffusion of iron, manganese, potassium, and zinc in parboiled rice kernels. LWT-Food Science Technology. 2016;**71**:138-148

[36] Hotz C, Kabir KA, Dipti SS, Arsenault JE, Bipul M. Rice fortification with zinc during parboiling may improve the adequacy of zinc intakes in Bagladesh. Journal of the Science of Food and Agriculture. 2015;**95**:379-385

[37] Jawahar S, Vaiyapuri V. Effect of sulphur and silicon fertilization on yield, nutrient uptake and economics of rice. International Research Journal of Chemistry. 2013;**3**:35

[38] Sharma A, Shankhdhar D, Shankhdhar SC. Enhancing grain iron content of rice by the application of plant growth promoting rhizobacteria. Plant, Soil and Environment. 2013;**59**:89-94

[39] Rana A, Joshi M, Prasanna R, Shivay YS, Lata N. Biofortification of wheat through inoculation of plant growth promoting rhizobacteria and cyanobacteria. European Journal of Soil Biology. 2012;**50**:118-126

[40] De Santiago A, Quintero JM, Aviles M, Delgado A. Effect of Trichoderma asperellum strain T34 on iron, copper, manganese, and zinc uptake by wheat grown on a calcareous medium. Plant and Soil. 2011;**342**:97-104

[41] Tariq M, Hameed S, Malik KA, Hafeez FY. Plant root associated bacteria for zinc mobilization in rice. Pakistan Journal of Botany. 2007;**39**:245-253

[42] Khan AG. Role of soil microbes in the rhizospheres of plants growing on trace metal contaminated soils in phytoremediation. Journal of Trace Elements in Medicine and Biology. 2005;**18**:355-364

[43] Clark RB, Zeto SK. Mineral acquisition by arbuscular mycorrhizal plants. Journal of Plant Nutrition. 2000;**23**:867-902

[44] Karthikeyan N, Prasanna R, Nain L, Kaushik BD. Evaluating the potential of plant growth promoting cyanobacteria as inoculants for wheat. European Journal of Soil Biology. 2007;**4**:23-30

[45] Karlen DL, Varvel GE, Bullock DG, Cruse RM. Crop rotations for the 21st century. Advances in Agronomy. 1994;**53**:1-45

[46] Kumar A, Yadav DS. Use of organic manure and fertilizer in rice (Oryza sativa)-wheat (Triticum aestivum) cropping system for sustainability. India Journal of Agriculture Science. 1995;**65**:703-707

[47] HarvestPlus. Biofortification Prioritization Index (BPI). Washington, DC: International Food Policy Research Institute (IFPRI); 2013. Available from: http://www.harvestplus.org/content/worldfood-day-new-ranking-toolguideinvestment-biofortified-crops-launched

[48] Garg M, Sharma N, Sharma S, Kapoor P, Kumar A, Chunduri V, et al. Biofortified crops generated by breeding, agronomy, and transgenic approaches are improving lives of millions of people around the world. Frontiers in Nutrition. 2018

[49] Yadava DK, Hossain F, Choudhury PR, Kumar D, Singh AK. Biofortification of crops: A sustainable tool for nutritional security in India. Indian Farming. 2018;**68**:37-42

[50] Finkelstein JL, Mehta S, Udipi SA, Ghugre PS, Luna SV, Wenger MJ, et al. A randomized trial of iron-biofortified pearl millet in school children in India. The Journal of Nutrition. 2015;**145**:76-81

[51] Wallace GA, Wallace A. Micronutrient uptake by leaves from foliar sprays of EDTA chelated metals. In: Nelson SD, editor. Iron Nutrition and Interactions in Plants. Basel: Marcel Dekker; 1982, 1982. pp. 975-978

Chapter 2

Zinc Biofortification in Rice (*Oryza sativa* L.)

Anjali Singh

Abstract

Rice is a major energy source and micronutrient for more than half of the world's population, but it lacks enough zinc to meet human nutritional needs. In addition, climate change, especially rising carbon dioxide levels in the atmosphere, is reducing zinc levels in grains. Therefore, rice bio-enrichment has been identified as a major goal for increasing zinc levels in grains to alleviate global zinc deficiency. There is a wide range to accelerate the development of High Zn varieties by applying biotechnology tools such as gene gun method and advanced genomics technology Successful intake and consumption requires an effective rice value chain, quality control and promotion of Zn bio-enriched varieties. Low zinc uptake, transport, and grain load have been identified as major bottlenecks in rice zinc bio enrichment. Nevertheless, the environmental and genetic factors of grain zinc accumulation in rice have not been fully studied. This review critically examines important genetic, physiological, and environmental variables that affect zinc uptake, transport, and utilization in rice. It also studies the genetic diversity of rice germ plasms and provides new genetic tools for bio-enhancement of zinc.

Keywords: rice, malnutrition, biofortification, zinc, biotechnology

1. Introduction

Biofortification is the process of increasing the nutrient content of a plant from seed to harvest. This is different from food fortification, which increases the nutritional value of edible crops during processing. Biofortification improves the nutritional value of crops during the plant's growth stage by embedding micronutrient content in growing crops. Bioenhancement of crops can be achieved by breeding or genetic engineering. In India, this is only done by breeding. Iron, zinc, and vitamin A deficiencies are the focus of bioenhanced research. These are micronutrients that affect most people around the world. In India, pearl millet (iron), wheat (zinc), sorghum (zinc), rice (zinc), sage (iron) and lentils are the main products (iron and zinc). Currently, bio-enriched pearl millet, rice and wheat are available to Indian farmers.

Biofortification, on the other hand, avoids the other three options by focusing on the production of nutrient-rich crops that can be cultivated and propagated using current agricultural practices. Fortification requires the use of unnatural additives,

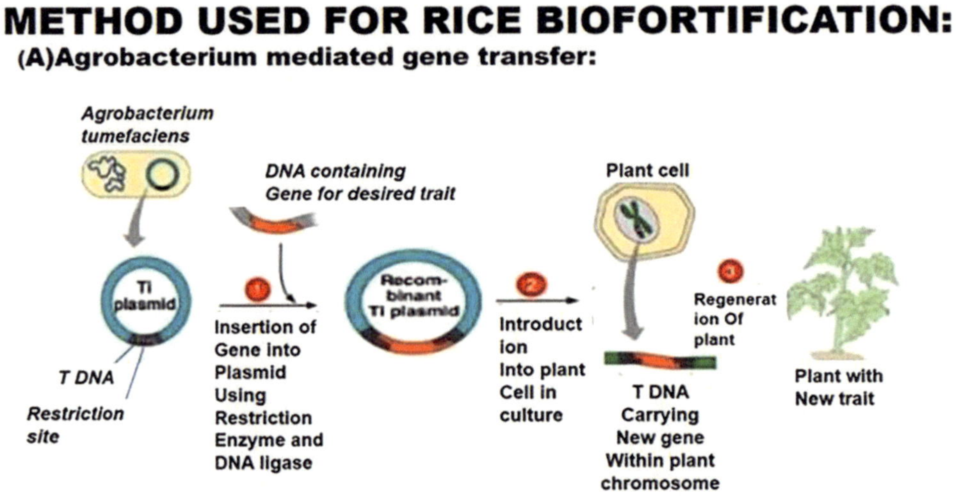

Figure 1.
Transformation of plants via A. tumefaciens.

but bio-fortification depends on the natural ability of the plant to produce and store nutrients. Agricultural bioenrichment, also known as bioenrichment, is the use of inorganic fertilizers to improve mineral element concentrations and/or mineral element mobility and soil solubility of edible plant parts. Agricultural bioenhancement is simple and inexpensive, but requires special attention to nutrient supply, application, and environmental impact. Biofortification is the process of improving plant species to increase the nutritional value of the ingested product. The following are some of the key approaches that can be taken with the Plant Breeding Initiative to increase the nutrient content of the foods produced. Rice breeding for higher grain Fe and Zn content, rice breeding for higher grain carotene content, rice breeding for higher grain folic acid content, plant transformation techniques, protoplast transformation, fine particles Transformation with a gun (or microscopic gun), transformation with Agrobacterium tumefaciens, identification and in vitro-transformation tissue selection, gene expression and regulation, protein expression, proteomics. For vitamins in the Fe, Zn, and B complexes, it is necessary to confirm the occurrence of variability among rice genotypes. Therefore, it is possible to select these materials within the breeding program. Simple selection of these superior genotypes in terms of nutritional value, albeit with traditional breeding techniques, may benefit the consumption of rice by the human population (**Figure 1**) [1].

2. Benefits of biofortification

In India, a movement similar to the Green Revolution aimed to end hunger. The country is expanding its edible grain production and is now almost self-sufficient as a result of the Green Revolution. The government is implementing several programs and measures to ensure that the population is consuming enough calories. However, the current focus is on improving the nutrient content of the diet. Many people do not get enough nutrients from their food intake, even if they have "sufficient food." The result is a problem called "hidden hunger."

3. Zn Deficiency

Human health problems are caused by zinc deficiency. More than half of the world's population suffers from zinc deficiency. Food fortification improves the trace element content of nutrients and improves nutrition and bioavailability of trace elements. Fortification has made progress in controlling micronutrient deficiencies, but new approaches are needed, especially in developing countries. The process of increasing the natural levels of biologically available nutrients in plants is called bioavailability. Bio-enriched crops are a cost-effective strategy for combating micronutrient deficiencies. Zinc deficiency in plants can be quickly addressed with a variety of effective fertilization techniques.

3.1 Zinc estimation

Seed samples were peeled and polished according to a standardized protocol for analyzing samples with XRF. Seeds were manually washed and peeled using a Harvest Plus-sponsored JLGJ4.5 non-ferrous huller (Jingjian Huayuan International Trade Co., Ltd., Jiangsu Sheng, China). The peeled brown rice was sieved to remove the broken rice grains, and the complete brown rice grains were cleaned with soft tissue paper. With the hands of a trained person, the sample was lightly rubbed on paper for 1 minute with the hands of a gloved person to ensure that non-rice particles were removed. Each brown rice sample was ground on a specially designed K710 non-ferrous rice mill (Krishi International India Ltd., Hyderabad, India) and the ratio of milled rice was calculated based on the weight of white rice to brown rice. Brown rice samples were polished for 90–120 seconds until the desired whiteness was reached, as white rice is the most common method of consumption. White rice was polished in the same way as brown rice. The time between polishing and washing has been reduced to prevent the bran particles from adhering to the polished grain. Using Energy Dispersive XRF (EDXRF) (OXFORD Instruments XSupreme 8000, Highwycombe Bucks, England), each sample of white rice or white rice (5 g) was analyzed on a Harvest Plus-funded IIRR. The fluorescence intensity of each sample was converted to zinc content (mg/kg) by scanning the sample using a pre-standardized method. Use the zinc concentration of brown rice and white rice [1].

Zinc is an important trace element for normal and healthy plant growth and reproduction. Zinc is classified as a micronutrient due to its low levels of 5100 mg kg1 in plant tissue. Iron, copper, zinc, manganese, cobalt, chromium, iodine and selenium are all important micronutrients in the food chain. Many plant enzymes, including functional, structural, and regulatory enzymes, rely on zinc for their activity. Zinc is also beneficial in plant glucose metabolism, sucrose and starch production, protein metabolism, membrane integrity, and auxin metabolism.

Zinc plays an important role not only in the development of the human immune system, but also in the cellular processes of all living organisms. The recommended daily zinc intake for adults is 15 mg. Zinc acts as a catalyst or structural component of various human and plant enzymes. Zinc is required for many essential enzymes such as RNA polymerase, superoxide dismutase, lactate dehydrogenase, alkaline phosphatase, aldolase, and phospholipase. Zinc deficiency can impair the development of embryos, fetuses, newborns and young children, impair the immune system and delay cell recovery. Zinc deficiency is said to be responsible for the prevalence and disability of children under the age of 5 in low-income countries. Zinc supplementation has been shown to reduce visible diarrhea and respiratory illness in humans. Inadequate intake of zinc in the human diet due to zinc deficiency. Zinc deficiency leads to dire conditions such as hair loss, memory loss, and weakness of the body's muscle tissue that occur in humans. Studies show that adult men need 11 mg of zinc daily and girls need 9 mg. During pregnancy, women should take 1315 mg of zinc daily. 3 mg of zinc per day for babies aged 7 months to 3 years, 5 mg per day for 48-year-old children, 8 mg per day for 913-year-old children Zinc deficiency stores zinc in the shell Because of this, it is common in people who are given grain. Grains are processed into flour. Foods rich in zinc include beef, chicken, almonds, walnuts, oatmeal, yogurt, cheese and milk. The bioavailability of the zinc and iron elements is significantly reduced due to the high content of phytic acid, dietary fiber and tannins, especially in cereal group plants.

It is a term used to refer to the fortification of foods, which increases the content of foods, especially trace elements, thereby improving nutrition and the bioavailability of trace elements. Adding iodine to table salt or adding Fe, Zn, and folic acid to bread crumbs is an example of fortification. The stability of the additive is low, which is a disadvantage in these applications. For example, adding folic acid to rice makes it easier to melt at high temperatures, and cooking rice makes it completely melt. Another drawback is that chemicals can compromise the quality of foods that are mixed in the long term. For example, additives containing iron oxidize and decompose over time, degrading the taste. Applying zinc to soil and/or crop leaves is a rapid way to increase zinc content (**Table 1**). Despite the fact that applying zinc to crops can increase yields and trace element concentrations, many farmers around the world (especially poor countries) do not. In agricultural bio-enhancement (soil/foliar fertilization, etc.), lowering the phytic acid/zinc molar ratio by reducing the phytic acid content of the crop can increase human Zn absorption.

A plant's ability to transport amino acids is critical. Both phloem and xylem are involved in amino acid translocation. As a result, amino acid translocation aids nitrogen recycling between roots and shoots and speeds up the plant's translocation of immobile nutrients, such as Zn. Furthermore, foliar application of urea to zinc fertiliser increases zinc transport throughout plants. Zinc applications will be a separate approach in the soil application of zinc, taking into account the growth periods of the plants.

Crops	Methods	References
Rice (*Oryza Sativa*)	Soil, Soil+Foliar	Özcan [2], Özcan et al. [3], Özcan [4], Phuphong et al. [5], Grija Veni et al. [6]

Table 1.
Different methods used for zinc biofortification in rice crops.

3.2 Zinc biofortified rice

Rice is the most important food crop in the world and is trusted by more than half of the world's population. Asia produces and consumes more than 90% of the world's rice. India is the second largest rice producer in the world, producing 112.76 million tonnes from 2017 to 2018. Deficiencies or accumulation of important amino acids, micronutrients and vitamins result from an imbalanced supply that alters human metabolism.

According to the World Health Organization, zinc deficiency is the fifth most common cause of illness in developing countries and the eleventh in the world. Globally, the prevalence of zinc deficiency in soil is estimated to be 20%. Zinc deficiency causes diarrhea and respiratory illness, killing 400,000 people worldwide each year. Zinc deficiency is also associated with poor growth, loss of appetite, skin lesions, taste bud disorders, delayed wound healing, hypogonadism, delayed sexual maturation, and impaired immune response. In India, zinc malnutrition causes 1.31 million disability-adjusted life years (DALYs) to be lost each year. Preliminary analysis of rice zinc bio-enhancements in India shows that of the 1.31 million DALYs lost, 0.142 and 456,000 DALYs under pessimistic and optimistic assumptions when zinc bio-enhanced rice is introduced. I found that I could save money. As a "public good of the world", the International Rice Research Institute and the International Agricultural Research in the form of the Wheat and Corn Improvement Center merged to form the International Agricultural Research Council Group (CGIAR), first carried out and led. it was done. The "Green Revolution" of the 1960s succeeded in improving grain production through the development of high-yielding varieties (HYV). However, HYV grains contain less nutrients. In the case of rice, milling further reduces nutritional levels, namely iron and zinc. Following this, in 1991, CGIAR responded to concerns expressed by the global nutrition community about micronutrient deficiency as a global issue, creating "micronutrient-rich" staple foods under signs of bioenhancement. Started research on. The Harvest Plus Challenge Program was launched by CGIAR in 2003 as a global program aimed at producing bio-enhanced staple crops such as wheat, rice, corn and cassava through plant breeding. Bioenhancement of rice grains with iron and zinc began in 1992 and 1995, respectively.

4. Techniques of biofortification

The major techniques or methods by which crops can be biofortified are mentioned below.

1. Agricultural practices: This involves using fertilizers to increase the amount of micronutrients in plants grown in soils that are deficient in those nutrients.

2. Traditional plant breeding: This is to create sufficient genetic variation in crop plants using traditional breeding methods, for example, to achieve high content of B. micronutrients. It involves mating types over several generations to produce nutritious plants and other desirable traits. In India, bio-enriched plants are produced only using this technology.

3. Genetic engineering/engineering: This involves adding DNA to the organism's genome to create new or altered traits, such as traits. B. Disease resistance to be introduced (**Figures 2** and **3**).

4.1 Technology efficacy of zinc biofortification in rice

The current zinc intake from common rice varieties was determined based on the per capita consumption of 220 grams of rice per day. Improved zinc intake from bio-enriched rice varieties was calculated assuming that India's current rice consumption pattern is maintained and bio-enriched rice varieties are being adopted as a technology.

Figure 2.
Managing zinc deficiency through agronomic approach.

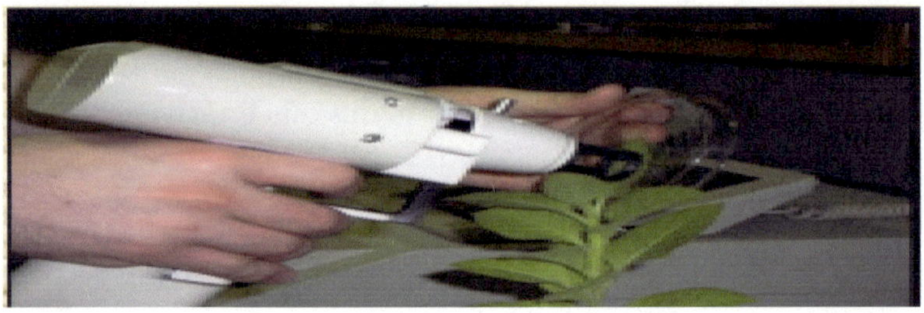

Figure 3.
Managing zinc deficiency through Particle Gun method.

Zinc is a component that contains more than 300 enzymes to repair cell damage, maintain fertility, synthesize proteins, and boost immunity among many important functions in human health [7]. Symptoms of zinc deficiency, large or small, can cause stunting, eczema, hair loss, delayed sexual maturation, and mental illness. Sustainable supply of dietary supplements (fertilization) and bioenrichment requires urgent efforts to overcome micronutrient deficiencies. Provision of supplementation through zinc fertilization of rice can increase zinc levels in rice [8]. However, this method is less effective due to nutritional loss from runoff, leaching, and the evaporation process. Therefore, a new strategy to overcome malnutrition of micronutrient is by the mean of biofortification. Biofortification provides a costeffective and sustainable solution in tackling the lack of nutrients supply [9]. This method is one of the plant breeding strategies to increase the zinc content in rice while improving the nutrition capacity with relatively low cost. Breeding materials are conventionally formed (hybridization and selection) or nonconventional (another culture and gene transformation). High zinc levels rice produced by these, inexpensive cost, production can be consumed directly by the middle to lower community as a source of energy and sources of nutrients [10]. As a result of these findings, rice has a high micronutrient content. Rice with a high micronutrient content can help customers get more micronutrients and overcome micronutrient insufficiency.

5. Biofortification challenges

Some of the challenges faced in biofortification and introducing biofortified food grains as part of the daily diet in India are discussed below.

- Due to the colour changes in the grain, people hesitate to accept biofortified food as in the case of golden rice.
- Farmers also should adopt this on a large scale.
- The initial costs also could be a barrier for people to implement.

Zinc deficiency is a serious problem in developing countries where white rice is the staple food. The creation of bio-enriched rice varieties in India was sought with the help of Harvest Plus, the Biotechnology Department, and the Indian Agricultural Research Council due to the high genetic diversity of white rice's high zinc content. Through the All-India Collaborative Rice Improvement Project (AICRIP), the Indian Rice Research Institute (IIRR) has enabled the release of rice varieties and is supporting India's rice bioenhancement program. Different sets of germ plasms from several national institutions have been characterized for zinc content in IIRR of brown rice using energy dispersive X-ray fluorescence spectroscopy. This indicates that the zinc range is 7.3–52.7 mg/kg. Assessment of zinc content in various wild germ plasm mapping populations, native varieties, and cultivars demonstrated the feasibility of favorable rebinding of high zinc and high yields. Nine genotypes from genetic resources and 344 strains from the mapping population showed zinc levels of _28 mg/kg in white rice, meeting the target zinc levels set by Harvest Plus. Through AICRIP biofortification trial constituted since 2013, 170 test entries were nominated by various national institutions until 2017, and four biofortified rice varieties were released. Only the test entry with target zinc content, yield, and quality parameters is promoted to the next year; thus, each test entry is evaluated for 3 years across 17 to 27 locations for their performance. Multilocation studies of two mapping populations and AICRIP biofortification trials indicated the zinc content to be highly influenced by environment. The bioavailability of a released biofortified rice variety, viz., DRR Dhan 45 was found to twice that of control IR64. The four released varieties generated through traditional breeding had technology efficacy ranging from 48 to 75 percent, with zinc consumption of 38 to 47 percent and 46 to 57 percent of the RDA for male and female, respectively. The results of germplasm characterization and population mapping for zinc content, as well as the construction of a national evaluation system for the release of biofortified rice varieties, have been reviewed in the context of the five biofortification programme criteria.

6. Conclusions

Bio-enrichment of zinc-enriched rice is an effective means of combating zinc malnutrition in rice-dominated developing countries. Some progress has been made in understanding the molecular basis of zinc accumulation. By developing bio-enriched rice varieties with high zinc content in white rice, with domestic and international funding in India, we are addressing zinc deficiency, especially in developing countries where rice is an important staple food. .. Plants are at the top of the food chain and produce large amounts of nutrients for consumption by other species. As a result, enhancing the uptake of minerals from the soil and increasing their mobility and bioavailability in the edible parts of plants benefits human and animal nutrition. In addition, future bioenhancement will be needed to fully understand the number of nutrients in soil and plant ecosystems. This is a potential means of significantly impacting the nutritional problems of human health and providing more nutrients to the world's population.

Acknowledgements

I would first like to thank my supervisor, Assistant Professor Dr. O.P.Verma, whose expertise was invaluable in formulating the research questions and methodology. Your insightful feedback pushed me to sharpen my thinking and brought my work to a higher level. In addition, I would like to thank my parents and my husband for their wise counsel and sympathetic ear. You are always there for me, who provided stimulating discussions as well as happy distractions to rest my mind outside of my research.

Conflict of interest

No conflict of interest.

Author details

Anjali Singh
Department of Genetics and Plant breeding, Acharya Narendra Deva University of Agriculture and Technology, Kumarganj, Ayodhya, Uttar Pradesh, India

*Address all correspondence to: anjalisinghgpb@gmail.com

References

[1] Sanjeeva Rao D, Neeraja CN, Madhu Babu P, Nirmala B, Suman K, Rao LVS, et al. Zinc biofortified rice varieties: Challenges, possibilities, and progress in India. Frontiers in Nutrition. 2020;7:26. DOI: 10.3389/fnut.2020.00026

[2] Özcan H, Taban S. Effect of zinc application on yield and grain zinc, phosphorus and phytic acid concentration of some rice genotypes. Journal of Soil Water. 2012;**1**(1):7-14

[3] Özcan H, Taban S, Tunaboylu ÖK, Çikili Y, Taban N. The effects of zinc application on the growth and zinc content of rice varieties having different kernel weights. Journal of Soil Water. 2013;**2**(2)

[4] Özcan H, Taban S. The effect of zinc application on agronomic parameters of some rice genotypes. Journal of Soil Science and Plant Nutrition. 2018;**6**(1):12-18

[5] Phuphong P, Cakmak I, Yazici A, Rerkasem B, Prom u Thai C. Shoot and root growth of rice seedlings as affected by soil and foliar zinc applications. Journal of Plant Nutrition. 2020;**43**(9):1259-1267

[6] Girija Veni V, Datta SP, Rattan RK, Meena MC, Singh AK, et al. Effect of variability of zinc on enhancement of zinc density in basmati rice grain grown in three different soils in India. Journal of Plant Nutrition. 2020;**43**(5):709-724

[7] Mares-Perlman JA, Subar A, Block G, Greger JL, Luby MH. Zink intake and sources in the US adult population: 1976-1980. Journal of the American College of Nutrition. 1995;**14**:349-357

[8] Welch RM. Effects of nutrient deficiencies on seed production and quality. Advanced Plant Nutrition. 1986;**2**:205-247

[9] Bouis H. Proceedings of The 1st International Conference on 'Rice for The Future'. Bangkok, Thailand: Kasetsart University; 2004. pp. 43-64

[10] Indrasari SD, et al. 2004. Indonesian Final Report Year III. Breeding for Iron Dense Rice: a Low Cost, Sustainable Approach to Reducing Anemia in Asia International Food Policy Research Institute (IFPRI) and Indonesian Center for Food Crops Research and Development (ICFORD) (nutrition aspect) (unpublished)

Chapter 3

Plant-Bacterial Symbiosis: An Ecologically Sustainable Agriculture Production Alternative to Chemical Fertilizers

Tuba Arjumend, Ercüment Osman Sarıhan and Mehmet Uğur Yıldırım

Abstract

Fertilizers have become a necessity in plant production to fulfill the rapid rise in population and, as a result, the increased nutritional needs. However, the unintended and excessive use of chemical fertilizers causes many problems and has a negative impact on agricultural production in many countries today. The inability to determine the amount, types, and application periods of the applied fertilizers adversely affects the natural environment, resulting in global warming and climate change, as well as the occurrence of additional abiotic stressors that have an impact on agricultural productivity. Hence, alternatives to chemical fertilizers and pesticides, such as the use of biofertilizers, must be explored for the betterment of agricultural production in a manner that does not jeopardize the ecological balance. Bacteria residing in the plant's rhizosphere can help with plant development, disease management, harmful chemical removal, and nutrient absorption. Introducing such phytomicrobiome into the agricultural industry is an effective approach as a result of its long-term and environmentally favorable mechanisms to preserve plant health and quality. Hence, this chapter aims at highlighting the deleterious effects of chemical fertilizers and providing a striking demonstration of how effectively plant-growth-promoting rhizobacteria (PGPR) can be used to increase the agriculture production in the context of climate change.

Keywords: chemical fertilizers and pesticides, climate change, global warming, biostimulants, phytomicrobiome, PGPR

1. Introduction

Today's global population is estimated to be over 7.9 billion people, which is expected to reach 9.9 billion in 2050, 34% higher than it is now [1]. Developing countries will account for nearly all of this overpopulation. To feed this growing population, agricultural lands must be used considerably more effectively, and production should be boosted by 70% compared to today's values [2]. Besides, agricultural production areas are unfortunately facing major ecological challenges, owing to human misapplications, natural calamities, as well as the impact of global climate change [3]. As a result of these factors, today the condition of our current lands is deteriorating leaving us with no choice but to grow nutrient-rich, chemical-free agricultural produce for human and animal use while using far less water and arable land than in the past. This is why a focus on both quality and quantity should be placed on food production without depleting natural resources. Developing and disseminating improved agricultural methods and technologies are equally critical.

Since cultivation areas are dwindling year after year, fertilizer mineral is a world market item that is vital to produce a higher plant yield per unit area and attain food security. It must be available in adequate quantities and in the proper balance to close the gap between nutrient supply from soil and organic sources and nutrient demand for optimal crop development [4]. Not just that, fertilizer is critical for the nation's economy to grow, as agriculture is the primary source of employment. By 2025, it will ensure food security for more than 8 billion people around the globe [5]. The increase in the use of chemical fertilizers by approximately 5 million tons in 10 years is a situation that should be considered while the agricultural areas are decreasing. However, it is more necessary to keep the soil's plant nutritional balance by considering climate, soil, and plant characteristics rather than the amount of chemical fertilizers utilized, and fertilizing based on soil analysis is critical.

2. The use of chemical fertilizers in agriculture

Fertilizer is recognized as one of the most valuable agricultural production inputs, and synthetic fertilizers are becoming increasingly popular around the world. The global fertilizer market was valued at $155.8 billion in 2019, with a compound annual growth rate (CAGR) of 3.8% predicted for the forecast period (2019–2024) [6]. Fertilizer consumption climbed from 10,777,779 million tons in 2015 to 14,495,815 million tons in 2020, a record high. The total global demand for fertilizers (N + P + K) was estimated at 198.2 million metric tons (mmt) in 2020/2021, according to the International Fertilizer Association (IFA). This was nearly 10 mmt, or 5.2% higher than in 2019–2020 and was the highest rise since the 2010–2011 fiscal year. Nitrogen experienced a 4.1% increase in demand to 110 mmt. Phosphorus demand increased by 7.0% (3.3 Mt), reaching 49.6 Mt., while demand for potash rose by 6.2% (2.2 Mt) to 38.5 Mt. [7]. In the last 50 years, the amount of chemical fertilizer used throughout the world has increased dramatically (**Figure 1**) [8].

Chemical fertilizers have also become more popular in Turkey in recent years, where the cultivation areas are decreasing every year, the need for fertilization is increasing, since more plant production per unit area is required. According to TUIK (Turkish Statistical Institute) 2021 statistics, both the use of fertilizers and nitrogen fertilizers has increased in agricultural production in Turkey in the last 10 years. TUIK statistics showed that annual fertilizer use in Turkey increased from 9,074,308 tons to 14,495,815 tons between 2010 and 2020, and the use of nitrogenous fertilizers increased from 5,995,500 tons to 9,774,691 tons within these values. The amount of fertilizer per unit production area is 107 kg/ha. The use of chemical fertilizers in agricultural inputs accounts for a share of 15–20% [9].

Advances in fertilization and agricultural applications have led to a significant increase in crop productivity in many regions, including Turkey. The most important chemical fertilizers applied to obtain more efficiency in plant production are those

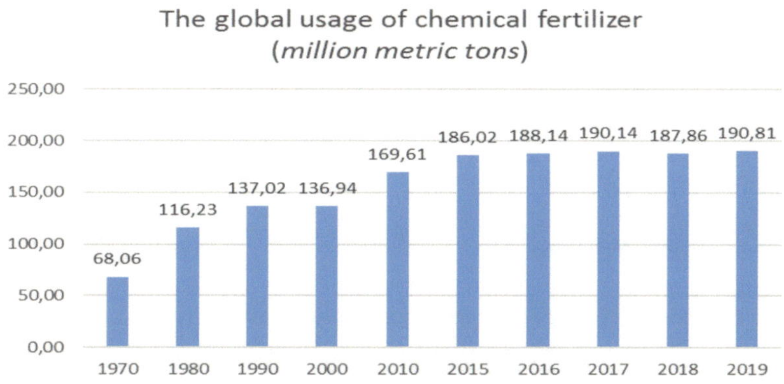

Figure 1.
Global usage of chemical fertilizer since 1970 [8].

containing nitrogen, phosphorus, and potassium. Nitrogen fertilizers (N), however, are the most widely used chemical fertilizers in the world, as well as in Turkey, and play a unique role in plant production. Potassium fertilizers (K_2O) are the second most consumed after nitrogen, followed by phosphorus fertilizers (P_2O_5) [8].

It has been determined that 87% of agricultural lands in Turkey have poor organic matter content [10]. Therefore, agricultural production is supported by fertilization, and nitrogen fertilizers constitute an important part of the total fertilizer applied. According to TUIK data, nitrogenous fertilizer usage rates as a percentage of total fertilizer use have shifted between 65 and 69% in the last 10 years [9]. Fertilizer use benefits plants in a variety of ways, including being a less expensive source of nutrients, having significant nutrient content and solubility, making it easily available to plants, and requiring less fertilizer, hence making it more suited than organic fertilizer [11]. Despite these advantages, mineral fertilizer has a number of negative environmental consequences as a result of rising consumption and decreased nutrient utilization efficiency. As a result, in intensive agricultural production systems, integrating intense cultivation with high nutrient utilization efficiency is a key difficulty.

3. Harmful effects of unnecessary chemical fertilizer use

Though conscious fertilization is desirable, the use of improper fertilizers can be extremely harmful, posing severe problems for current and future generations [12]. Sometimes, unfortunately, a wrong perception occurs among the producers of chemical fertilization. It is thought that more efficiency can be obtained by using more chemical fertilizers. Contrary to popular belief, the "LAW OF DECREASING PRODUCTION" is valid in fertilization. That is, the benefit derived from fertilization rises up to a point, after which continuing to apply fertilizer causes harm rather than a benefit.

The unintended and excessive use of chemical fertilizers to boost yields, as well as rising reliance on them, has a negative impact on the agricultural production system's sustainability as well as financial losses in many countries today [13]. Certain factors, such as changes in fertilizer type, variations in application times, the producer's lack of understanding in this area, and improper fertilizer applications, in particular, have been found to have quite substantial environmental consequences and threatening effects on the health and life of living creatures [14]. The inefficient and not demand-oriented fertilization applications in agricultural production can lead to soil acidity and soil crust, low organic matter and humus content, heavy metal accumulation, decrease in pH values, soil salinity, plant nutritional imbalances, limited plant growth, erosion, a decline in microbial activity and efficacy and emission of gasses containing substances that damage the atmosphere and the ozone layer, and eventually the greenhouse effect [15].

The issues at the forefront of the detrimental environmental effects of chemical fertilizers are highlighted here.

3.1 Increased acidity of the soil

Excessive soil acidity induced by fertilizers is a major cause of soil degradation across the world. Fertilizers, especially nitrogen, acidify soil when applied in excess. This scenario has negative consequences, such as the crops' incapacity to absorb phosphate, the proliferation of hazardous ion concentrations in the soil, hindrance of crop development, and suppression of microorganism activity [16]. If ammonium sulfate fertilizer is given to acidic soil, for example, the acidity level will become even higher. One-way ammonium sulfate fertilization of tea, according to research conducted in the Rize province of Turkey, considerably increased the acidity of low-pH soils. Currently, 85% of the land has fallen below pH 4, which is deemed critical. Likewise, in Nevsehir province, the pH of the soil has dropped to 2 as a result of nitrogen fertilization of potatoes grown in 100-fold increasing acidity over the last 25 years [17].

Hao et al. [18], carried out a field experiment to measure soil acidification rates in response to varied fertilizer sources and N rates, including control, optimal urea, conventional urea, optimized NH_4Cl, and conventional NH_4Cl plots, nitrogen addition resulted in average H^+ production of 4.0, 8.7, 11.4, 29.7, and 52.6 keq ha^{-1} $yr.^{-1}$, respectively. This was followed by a 1–10% decrease in soil base saturation and a 0.1–0.7 unit decrease in soil pH in the topsoil (0–20 cm). In a greenhouse study conducted to evaluate the effect of conventional nitrogen fertilizer on soil salinity and acidity, a significant rise in both soil acidity and salinity was witnessed as N input increased after one season, with pH decrease ranging from 0.45 to 1.06 units [19]. Moreover, after 21 years of application, chemical N fertilizer dropped the soil pH from 6.20 to 5.77, a 0.02 pH unit drop per year [20]. In another study, an evaluation of the impact of long-term fertilizing techniques on soil samples revealed a fall in soil pH from 8.4 to 7.5 [21]. Because nutrients are less available to plants in acidic soil, serious plant nutritional deficiencies are prevalent, resulting in overall crop reduction.

3.2 Deposition of heavy metals

Heavy metal deposition in soils is mostly caused by the manufacture and consumption of industrial products, although fertilizers and pesticides used in agriculture also contribute significantly. Arsenic (As), copper (Cu), nickel (Ni), cadmium (Cd), and uranium (Ur), among other heavy metals, can build up in the soil following repeated chemical fertilizer applications, particularly phosphorus (P) fertilizers and their source material [22–24]. These toxic heavy metals not only pollute the environment, but they may also cause soil degradation, plant development retardation, and perhaps impair human health through food chain contamination harming the central nervous system, circulatory system, excretory system, and cardiovascular system, as well as cause bone damage, endocrine disruption, and possibly cancer [25].

Phosphorus (P) fertilizer is widely utilized in agriculture due to its vital function in crop growth and production [26]. However, P fertilizer has been recognized as the predominant cause of HMs pollution in soil when compared to potassium (K) and nitrogen (N) fertilizers [27]. According to a 10-year field trial, P fertilization aided Zn, Pb, Cd, and As buildup in the topsoil. With increasing P application, the threshold cancer risk (TCR) associated with As and Cd increased [28]. Likewise, another experiment concluded that frequent application of P fertilizer and the extended residence period of HMs may generate a large accumulation of HMs in soils [29].

Heavy metals are concentrated in agricultural soil as a result of improper application of commercial fertilizers, manure, sewage, or sewage sludge [30]. The results of the study conducted by Huang and Jin [31] suggested that the long-term usage of exaggerated synthetic fertilizers and organic manures contributed to the heavy metals (HMs) accumulation in the soils. Research carried out by Atafar et al. [32], confirmed that the fertilizer use enhanced the amounts of Cd, Pb, and As in cultivated soils. Before fertilization, the Cd, As, and Pb concentrations in the studied location were 1.15–1.55, 1.58–11.55, and 1.6–6.05 mg/kg, respectively; after harvesting, values were 1.4–1.73, 26.4 5.89, and 2.75–12.85 mg/kg soil for Cd, As, and Pb, respectively. The findings of another study concluded that chemical fertilizer usage increased the availability of Cu, Ni, Pb, and Zn as well as the buildup of Cd, Cu, and Zn in the greenhouse soil [33].

3.3 Salinity of the soil

Salts are a common component of chemical fertilizers and are considered destructive to agriculture because they harm soil and plants. Increases in the salinity of the soil can be seen by natural or artificial means. Artificially induced salinity is the result of the accumulation of fertilizers used in large quantities over long periods of time in areas where intensive farming is practiced, making the soils unsuitable for production [22, 34, 35]. Following one season of conventional nitrogen fertilizer, electrolytic conductivity increased from 0.24 to 0.68 mS cm^{-1} [19]. Long-term intensive farming raised soil electrical conductivity (ECe), which rose from "low salinity" (1.5 dS m^{-1} 0.49) to "highly saline" (6.6 dS dS m^{-1} 1.35) levels [21].

Soil salinity is a major global issue that has a negative impact on agricultural output. Salinization of agricultural land diminishes economic advantages greatly, as demonstrated by Welle and Mauter [36] in California, where salinization lowered overall agricultural income by 7.9%.

3.4 Nutritional inadequacy

Inorganic fertilizers used recklessly can cause nutritional imbalances in the soil, thus limiting the intake of other essential nutrients. If the common NPK type is frequently used, secondary and micronutrient deficiencies occur in the soil and crop. Excess nitrogen and phosphate fertilizers, for instance, enable the plant to absorb more potassium than it requires. In acidic soils, lime and lime-containing fertilizers lead to the retention of micro plant nutrients, such as P, B, Fe, and Zn in the soil. Over-applied phosphorus fertilizers also prevent the uptake of nutrients, such as Ca, Zn, and Fe, and reduce their efficacy [22, 37].

3.5 The influence on soil friability

Soil compaction is a key component of the land degradation syndrome and a serious issue for modern agriculture, negatively impacting soil resources [38]. Overuse of fertilizers for extended periods of time and intensive cropping are two of the main causes of compaction. Chemical fertilizers damage soil particles, resulting in compacted soil with poor drainage and air circulation [39]. Reduced soil aeration has an impact on soil biodiversity. Microbial biomass may be diminished as a result of severe soil compaction. Soil compaction may not affect the amount of macrofauna, such as earthworms, but it does affect the distribution of macrofauna, which is important for soil structure.

Soil compaction leads to high soil strength and bulk density, poor drainage, poor aeration, limited root growth, erosion, runoff, and soil deterioration, hence resulting in low permeability, hydraulic conductivity, and groundwater recharge [40, 41]. High soil compaction stifles root growth, reducing the plant's ability to absorb nutrients and water. Compaction, according to reports, reduces root growth and yield by more than 80% [42]. As the soil bulk density increases, nitrification drops by 50%, and plants use less N, P, and Zn from the soil [43]. The findings of the research conducted by Massah and Azadegan [44] suggested that in non-compacted and severely compacted soils, bulk density increased from 1.34 to 1.80 $Mg.m^{-3}$, and penetration resistance increased from 0.89 to 3.54 MPa, respectively. Soil compaction reduced permeability by 81.4%, accessible water by 34%, and yields by 40%.

3.6 Soil structure and microbial activity deterioration

In agricultural production, the unintentional, excessive, or random application of chemical fertilizers and pesticides degrades the chemical, biological, and physical structure of the soil, resulting in a rise in pathogen and pest populations [45, 46]. Moreover, with intensive and unconscious chemical fertilizer applications, the amount of organic matter in the soil decreases, which adversely affects the microorganism activities and causes the structure of the soil to deteriorate. If the same fertilization errors are repeated, soil structures will deteriorate with each passing year, plant growth will slow as fertilizer doses are increased, and the overall amount of product obtained will decrease. Some of the fertilizers will not be able to hold on to the soil and will be removed with the water. The conversion of nutrients into forms that plants can benefit from will be reduced.

Soil microbial activity is a crucial component of soil health, and soil organisms serve as a mechanism for nutrient recovery, as well as provide a variety of other environmental functions. Chemical fertilizer misuse can have a detrimental and lethal effect on soil quality and microbial community structure, including earthworms, and other soil inhabitants. Prolonged consumption of chemical fertilizers can cause a significant drop in soil pH, which has been associated with a decrease in bacterial diversity and major changes in bacterial community composition [47]. Nitrogen usage in agriculture has a deleterious influence on the nitrogen cycle and the activities of related bacterial communities, including nitrogen-fixing microorganisms such as Rhizobium sp. [48]. Besides, excess nitrogen fertilizers limit the activities of nitrifying bacteria [49].

3.7 Contamination of water bodies and nitrate accumulation

It is critical to emphasize the importance of understanding how to apply chemical fertilizers properly. Chemical fertilizers, as part of their larger threat to the environment, animals, and human health, eventually leak into our water bodies, such as ponds, streams, and groundwater, contaminating water supplies, exposing humans and animals to a variety of short- and long-term hazardous chemical effects on their health and bodies. In ideal conditions, it is estimated that roughly 2–10% of fertilizers interfere with surface and groundwater [50]. The accumulation of nitrates in the water emerges as a result of the use of N fertilizers in the agricultural field, which is increasing day by day. Even under ideal conditions, only 50% of the nitrogen fertilizer given to the soil can be taken up by plants; 2–20% evaporates, 15–25% combines with organic compounds in the clay soil, and 2–10% is discharged into streams, rivers, and streams with surface runoff [50, 51]. Nitrate, a frequent contaminant of surface and groundwater, can cause serious health concerns, including inflammation of the colon, stomach, and urine systems. Furthermore, these compounds have been reported as carcinogens that can have a harmful impact on human health. They also have the potential to induce disorders in infants, such as methemoglobinemia, a condition in which the blood carrying capacity is limited due to a decrease in hemoglobin.

4. Agriculture and fertilizers' contribution to global warming and climate change

Though the rise in agricultural productivity alleviated poverty, it also posed a threat to the ecosystem due to its negative consequences. Rising levels of synthetic fertilizer application for agricultural production, for instance, increase greenhouse gas emissions, eroding the protective ozone layer, and exposing humans to harmful ultraviolet radiation [52]. Above all, agriculture is responsible for a major fraction of the greenhouse gas (GHG) emissions that are driving climate change, accounting for 17% directly from agriculture activities and another 7–14% through land-use changes.

During the production of nitrogenous fertilizer, greenhouse gases, such as CO_2, CH_4, and N_2O are released. Moreover, nitrous oxide emissions from soils, fertilizers, manure, and urine from grazing animals, as well as methane generation by ruminant animals and paddy rice agriculture, are the most significant direct agricultural GHG emissions. Both of these gases have a far larger potential for global warming than carbon dioxide.

Agriculture is the primary source of anthropogenic N_2O emissions, accounting for 60% of total emissions. It has a 310-fold greater global warming potential than carbon dioxide. Excess nitrogen fertilizer application results in nitrogen oxide emissions (NO, N_2O, NO_2), which cause serious air pollution [51]. The primary issue with nitrous oxide emissions is the impact of global warming and the function of nitrous oxides in ozone degradation, encouraging the decomposition of the ozone layer [53] and resulting in atmospheric "holes," exposing humans and animals to excessive UV radiation [54]. Water vapor, hydrogen sulfide, and chloro-fluoro hydrocarbons are among the other gases that contribute to ozone depletion [55].

After being volatilized or released from fertilized fields, ammonia is deposited in the atmosphere and oxidized to generate nitric and sulfuric acids, resulting in acid rain. Acid rain has the potential to harm flora, buildings, and species that live in lakes and reservoirs [56]. Methane emissions from transplanted paddy fields are also a major concern, as methane is a powerful greenhouse gas whose concentration is doubled when ammonium-based fertilizers are used. These gases all contribute to global warming and climate change [57].

Climate change is gaining traction, resulting in major global temperature spikes, as well as the prevalence of additional abiotic stressors that are reducing crop output.

Significant production losses in major grain crops have been attributed to climate change, resulting in 3.8% yield reductions for maize and 5.5% for the wheat [58, 59].

Fertilization, which is one of the most essential inputs in agricultural operations, increases productivity on the one hand, but its overuse might have negative consequences on the other. Excessive usage of agricultural chemicals jeopardizes the long-term viability of agriculture. Today, the fast expansion in agricultural productivity has begun to slow down [45, 56]. Clean food production becomes inevitable with a healthy and reliable agriculture system that does not require chemicals.

Given that chemical fertilization cannot be completely eliminated in agricultural applications, in this scenario, sustainability initiatives and the usage of ecologically sound technologies can help achieve the goal of enhancing healthier crop productivity whilst eliminating unnecessary input and thereby mitigating harsh weather conditions, as well as improving soil health by sequestering carbon and retaining organic material and mineral nutrients in the soil [60]. Hence, it is vital to use alternatives, such as Plant-Growth-Promoting Rhizobacteria (PGPR), to support sustainable agricultural productivity and everlasting soil fertility and to build production strategies that will aid in the proliferation of beneficial soil microorganisms activities.

5. Plant-growth-promoting rhizobacteria (PGPR): an ecologically sustainable alternative to chemical fertilizers for agricultural production

The rhizosphere is a well-defined ecological niche that consists of the volume of soil surrounding plant roots and is home to a wide range of microbial species [61, 62]. As a result of phytomicrobiome research, certain plant-microbe interactions that directly aid in plant nutrition are beginning to emerge [63]. Microbes have the power to positively influence plant growth and combat the majority of modern agriculture's challenges, making them a promising alternative for agricultural sustainability. The rhizomicrobiome is indispensable for agriculture because of the extensive diversity of root exudates and plant cell debris that attract diverse and unique patterns of microbial colonization. Fertilizer requirements are often lower in soils with dynamic microbial ecologies and rich organic matter than in traditionally treated soils [64].

Despite the fact that the rhizosphere is home to a diverse range of microbes, including bacteria, fungi, algae, protozoa, and actinomycetes, bacterial colonies are predominant [65, 66]. The bacterial community in the soil, in particular, has the potential to proliferate quickly and use a wide variety of nutrient sources. A group of natural soil bacterial flora that resides in the rhizosphere and grows in, on, or around plant roots [67] and has a beneficial effect on the plant's overall health is referred to as PGPR [68]. PGPR is a nonpathogenic, beneficial bacterium that promotes plant growth by modifying hormone levels and nutritional requirements, as well as reducing stress-related damage [69]. Nutrient absorption is thought to be increased as a result of the increased root surface area mediated by PGPR. Besides, they mineralize organic contaminants and are employed in polluted soil bioremediation [70]. When compared to other microorganisms, PGPR has unique characteristics, such as the ability to synthesize growth regulators, nitrogen fixation, phosphorus solubilization, siderophore generation, nutrients, and mineral solubilization, demonstrating their exceptional tendency in stimulating plant growth [71]. They are also environmentally friendly and ensure that nutrients from natural sources are available at all times. In addition to stimulating plant growth through their active mechanisms, the bacterial colonies in the rhizosphere have a considerable influence on suppressing phytopathogenic microorganisms. Beneficial rhizobacteria can emit antibiotics and other chemicals that are effective at inhibiting pathogens [72].

The fundamental impacts that rhizosphere bacteria have on plants have evolved into an important mechanism for protecting plant health in an environmentally sustainable manner [73]. They participate in a variety of biotic activities in the soil ecosystem to keep it active and productive for farming systems [74]. Furthermore, in recent times, PGPR has garnered much attention for its potential to substitute agrochemicals for plant growth and yield through multiple processes, including decomposition of organic matter, recycling of essential elements, soil structure formation, production of numerous plant growth regulators, degrading organic pollutants, stimulation of root growth, and solubilization of mineral nutrients, which are important for soil health [75]. It is cost-effective and environmentally beneficial to replace chemical fertilizers with PGPR, as well as to identify the most effective soil and crop management approaches in an attempt to develop more sustainable farming and soil conservation fertility [76]. The employment of phytomicrobiome representatives as a long-term disease prevention and nutrient supplement method in farming production might help to reduce the negative impacts of pesticide usage [77]. The inoculated plant's biocontrol and induction of disease resistance, biological N2 fixation, phosphate solubilization, and/or phytohormone synthesis are all potential explanations for PGPR's growth-promoting actions [78].

PGPR has both direct and indirect modes of action as a biofertilizer and a biopesticide.

6. The effect of PGPR on plant nutrient supplementation

6.1 PGPR as biofertilizers

One of the most prevalent ways for increasing agricultural production is to improve soil fertility. PGPR promotes soil fertilization through the biofixation and biosolubilization processes (**Figure 2**).

6.1.1 Biofixation of atmospheric nitrogen

Nitrogen (N) is found in all forms of life and is one of the most significant mineral nutrients for plant growth as it is a crucial component for various physiological activities in plants, including photosynthesis, nucleic acids, and protein synthesis [80]. Unfortunately, due to the low degree of reactivity, no plant species are capable of directly converting atmospheric dinitrogen into ammonia and using it for growth, hence making the plants dependent on biological nitrogen fixation (BNF). Nitrogen fertilizer, as being the most effective approach to nitrogen supplementation, has been an integral part of modern crop production and agricultural systems; yet, their continued and undesirable use is contaminating the climate. Though carbon dioxide (CO_2) is widely regarded as the primary cause of climate change, nitrous oxide (N_2O), which has a 265-fold higher heat-trapping efficiency than CO_2 [81], is indeed a significant contribution. PGPR in this regard is a potential alternative to minimize the fertilizer requirements to a certain degree as the majority of the plant microbial community can either directly fix atmospheric nitrogen through legume-rhizobium interaction or indirectly by helping nitrogen fixers via their secretion [82].

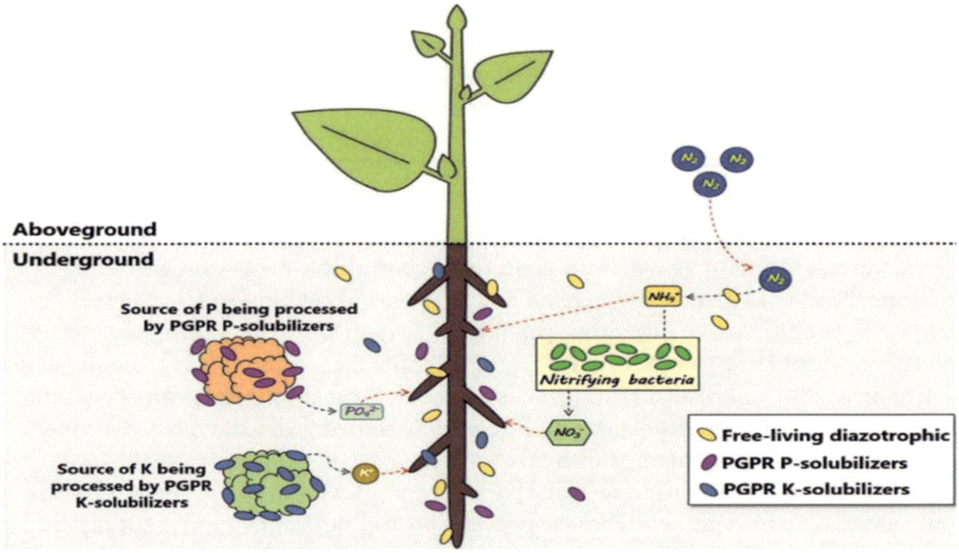

Figure 2.
PGPR's mechanism of action [79].

Worldwide, total N fixation is estimated to be ~175Tg, with symbiotic nitrogen fixation in legumes accounting for ~ 80 Tg by fixing 20–200 kg N year-1, while the remaining nearly half (~88 Tg) is industrially fixed during the production of N fertilizers [83]. The most prominent symbiotic nitrogen fixer is Rhizobium [84], whereas Azospirillum, Acetobacterdiazotrophicus, Azotobacter, Herbaspirillum, Cyanobacteria, Bacillus, Paenibacillus, Gluconacetobacter, and Azoarcus, etc., represent the free-living N fixers [85].

Symbiotic nitrogen fixation: A mutualistic association between a microorganism and a plant is known as symbiotic nitrogen-fixing. The N-fixing symbiosis between rhizobia and legumes is the most well-studied and utilized beneficial plant-bacteria interaction. In this interaction, legumes supply rhizobia with reduced carbon and a protected, anaerobic environment that is necessary for nitrogenase activity, while rhizobia feed legumes with biologically accessible nitrogen. The bacteria enter the root first, causing the growth of nodulation, which converts atmospheric nitrogen into usable forms (primarily NH3) [86]. Rhizobia can fix up to 200 kg of nitrogen ha – 1 by establishing symbiotic relationships with more than 70% of leguminous plants, thus making it available to plants.

Free-living nitrogen-fixing: Several nitrogen-fixing microorganisms do not interact in a symbiotic manner. These microorganisms are free-living and rely on plant leftovers or their own photosynthesis to exist. Although free-living nitrogen fixers do not enter the plant's tissues, a tight interaction is developed where these bacteria reside close enough to the root that the atmospheric nitrogen fixed by the bacteria is taken up by the plant, resulting in greater nitrogen absorption. Besides, other bacteria that do not fix nitrogen have been demonstrated to boost nitrogen uptake in plants, resulting in increased nitrogen use efficiency [87], most likely due to increased root development, which allows plants to reach more soil [63]. Evidence of PGPR involvement in the plant N budget has been identified for various plants, particularly sugarcane [88].

Rhizobial N-fixation is an integrated approach for disease control, growth stimulation, as well as providing and maintaining the nitrogen level in agricultural soils around the world, thus minimizing the need for extensive N-fertilizer application and limiting the soil and environmental challenges associated with it.

6.1.2 Phosphate solubilization

Phosphorus (P) is the most significant vital element in plant nutrition (N), alongside nitrogen [89]. It is involved in a number of major metabolic activities in plants, including macromolecular biosynthesis, photosynthesis, respiration, energy transfer, and signal transduction [90]. Although most soils hold a significant amount of phosphorus, which builds over time as a consequence of fertilizer treatments, plants have access to only a small portion of it. Despite the fact that P is abundant in both organic and inorganic forms in the soil, only 0.1% of it is available to plants because 95–99% of phosphate is either insoluble, immobilized, or precipitated [91]. Plants can absorb mono and dibasic phosphate on their own, but organic and insoluble phosphate must be mineralized or solubilized by microorganisms [92]. Phosphate anions are highly reactive and, depending on the soil quality, can be trapped by precipitation with cations including Mg^{2+}, Ca^{2+}, Al^{3+}, and Fe^{3+}. Plants cannot absorb phosphorus in these forms because it is highly insoluble. As a result, plants only get a small percentage of the total.

When deficient, phosphorus-based fertilizers are typically used to replenish soil P, which is readily available to plants. Supplementing P with commercial fertilizers, however, is not an ideal option due to their high cost and sometimes inaccessibility to plants since they are easily lost from the soil and subsequently mix with local streams, contaminating both terrestrial and aquatic environments [93]. Therefore, phosphorus solubilization, in addition to nitrogen fixation, is also vital biologically. Phosphate solubilization is among the most profound consequences of PGPR on plant nutrition. Persistent plant growth, PGPR, plays a major role in solubilizing phosphorus [94]. The potential of various bacterial species to solubilize insoluble inorganic phosphate compounds such as dicalcium phosphate, tricalcium phosphate, rock phosphate, and hydroxyapatite has been documented by many researchers. Phosphate can be dissolved in insoluble forms by a variety of PGPR, including Pseudomonas, Bacillus, and Rhizobium. PGPR solubilizes P by employing a number of mechanisms, including the synthesis of organic acids and extracellular enzymes, to make use of inaccessible forms of P, hence assisting in the availability of P for plant absorption. Miller et al. [95] pointed out two processes—acidification of the external medium via the release of low molecular weight organic acids (such as gluconic acid) that chelate phosphate-bound cations and the formation of phosphatases/phytases that hydrolyze organic forms of phosphate compounds. Phosphorus solubilizing bacteria (PSB) has been shown to lower the recommended P dose by approximately 25% [96] and is even more efficacious when combined with other PGPRs or mycorrhizal fungi, reducing the P supplementation to 50%. As a result, the risk of P runoff and eutrophication is mitigated [97].

6.1.3 Solubilization of potassium

Potassium (K) deficit has become a severe crop production bottleneck. Plants with insufficient potassium have poor root development, low seed production, a slow growth rate, and a decreased yield. Soluble potassium concentrations in soil are typically low; over 90% of the potassium in the soil is in the form of insoluble rocks and silicate minerals [98]. Several microbes, particularly fungal and bacterial genera, have close connections with plants and can release potassium in accessible form from potassium-bearing minerals in soils through the synthesis and secretion of organic acids [99–101]. Setiawati and Mutmainnah [101] synthesize organic acids produced by soil microorganisms, such as acetate, ferulic acid, oxalate, coumaric acid, and citrate, which significantly increase mineral dissolution rates and proton production by acidifying the soil rhizosphere and resulting in mineral K solubilization. As a result, using potassium-solubilizing PGPR as a biofertilizer in agricultural production can reduce agrochemical use while also encouraging environmentally friendly crop production.

6.1.4 Iron sequestration by siderophore production

Iron (Fe) is a major bulk mineral abundantly available on Earth, yet it is inaccessible in the soil for plants, owing to the fact that Fe^{3+} (ferric ion), the most common form of Fe found in nature, is hardly soluble [102]. PGPRs are the right fit to address this issue as they produce siderophores, which are tiny organic compounds that increase Fe absorption capability when it is scarce. Since PGPR can form siderophores, they are a valuable asset in supplying the plant with the necessary iron. Siderophores released by PGPRs boost plant growth and development via facilitating access to Fe in the soil surrounding the roots [103]. Plant growth can be stimulated directly by siderophore-producing bacteria, which improves plant Fe intake, or indirectly by suppressing the activities of plant pathogens in the rhizosphere, which limits their Fe availability [104]. Pathogen suppression is induced by the synthesis of siderophores, which decrease pathogen survival by chelating available Fe and therefore restricting pathogen survival [105]. In the presence of other metals, such as nickel and cadmium, a robust siderophore, such as the ferric-siderophore complex, is crucial for Fe uptake by plants [106]. Siderophores alleviate stress on the plants caused by potentially hazardous metals, such as Al, Cd, Cu, Pb, and Zn, found in polluted soil via forming stable compounds with them [107]. This phenomenon is beneficial for reducing plant stress induced by potentially harmful metals found in contaminated soils. Furthermore, siderophore-expressing rhizobacteria could be a potential alternative to chemical fertilizers by concurrently addressing salt-stress effects and Fe limitation in saline soils.

6.1.5 Exopolysaccharide synthesis or biofilm formation

One of the many advantages of rhizobacteria in encouraging plant growth and controlling plant diseases is their ability to synthesize polysaccharides. Multifunctional polysaccharides, for instance, structural polysaccharides, intracellular polysaccharides, and extracellular polysaccharides, are synthesized by specific bacteria. Exopolysaccharide production is critical for biofilm development, and root colonization can influence microbial interactions with root appendages. The colonization of plant roots by EPS-producing bacteria aids in the separation of free and insoluble phosphorus in soils, circulating critical nutrients to the plant for appropriate growth and development, as well as protecting it against disease attacks. EPS-producing bacteria have a variety of roles in plant-microbe interactions, including protection against desiccation, stress [108], adherence to surfaces, plant invasion, and plant defense response [109]. Plant exopolysaccharides produced by plant-growth-enhancing rhizobacteria are critical in stimulating plant growth because they act as an active signal molecule during beneficial interactions and generate a defense response during the infection phase [110]. Some plant-growth-promoting rhizobacteria that produce exopolysaccharides can also bind cations, including $Na+$, implying that they may play a role in limiting the amount of $Na +$ available for plant uptake and thereby reducing salt stress [111].

6.2 Production of biostimulants by PGPR

Phytohormones, commonly known as plant growth regulators, are organic chemicals that, at low levels (less than 1 mM), promote, inhibit, or modify plant growth and development [112]. Phytohormones are categorized based on where they act. Botanists recognize five main kinds of phytohormones: Auxins, Gibberellins, Ethylene, Cytokinins, Ethylene, and Abscisic acid.

Phytohormones stimulate root cell proliferation by overproducing lateral roots and root hairs, resulting in increased nutrition and water intake [113]. This is crucial for regulating nutrient uptake depending on soil composition and environmental circumstances. Slower primary root development and a spike in the proportion and length of lateral roots and root hairs are the most common effects.

Phytohormones play an important role in regulating developmental processes and signaling networks that are involved in plant adaptation to a variety of biotic and abiotic stressors [114]. Abiotic stressors, however, disrupt plant growth by altering endogenous levels of phytohormones [115]. Surprisingly, some bacteria, such as PGPR, may stimulate plants to produce phytohormones.

A diverse spectrum of rhizospheric microorganisms is capable of producing growth hormones that can promote cell proliferation in the root architecture by inducing an increase in nutrition and water intake by encouraging root hair growth, thus regulating overall plant growth and development, as well as activating pathogen defensive responses [116]. Important biological rhizobacteria can adjust to their surroundings and develop stress tolerance by repairing plant roots. The production of growth metabolites by PGPRs may help provide water stress resistance in host root colonization, resulting in higher optimal crop output.

Auxin is a critical molecule that regulates most plant functions, either directly or indirectly, and indole-3-acetic acid (IAA) is the most abundant and physiologically potent phytohormone that regulates gene expression by upregulating and downregulating it [116, 117]. More than 80% of rhizospheric bacteria have been known to be capable of synthesizing and releasing auxins. IAA produced by PGPR regulates a wide range of processes in plant development and growth, including cell division, differentiation, organogenesis, tropic responses, primary root elongation, and the formation of lateral roots [118]. As a result of the increased root surface area and length mediated by bacterial IAA, plants have better access to soil nutrients. Under salinity stress circumstances, the secretion of IAA by PGPR may have a key function in managing and regulating IAA concentrations in the root system, resulting in improved plant salinity stress responses [119]. Besides, microbe-induced IAA can boost root and shoot biomass output in water-stressed situations [120].

Gibberellins (GA) are another type of phytohormone produced by rhizobacteria. Different activities in higher plants, such as seed germination, root and leaf meristem size, cell division and stem elongation, floral induction, fruiting, and the flowering process, growth of the hypocotyl and stem, are all mediated by GA [121]. However, shoot elongation is by far the most significant physiological function of GA [122], which modifies the morphology of plants.

Cytokinins are a type of growth regulators that are responsible for seed germination, production of shoots, vascular cambium sensitivity, the proliferation of root hairs, improvement of cell division and root development, interactions of plants with pathogens, and nutrient mobilization and assimilation [123, 124], but suppress root elongation and lateral root formation [125, 126]. They are especially important for the cell cycle's progression. Cytokinin, either alone or in combination with other phytohormones like abscisic acid and auxin, can help salt-stressed plants grow faster while also improving resistance by altering the expression of genes [127]. PGPRs, such as Bradyrhizobiumjaponicum, Azospirillumbrasilense, Pseudomonas fluorescens, Arthrobactergiacomelloi, Paenibacilluspolymyxa, and Bacillus licheniformis, have been demonstrated to produce cytokinin (particularly zeatin) [69]. Cytokinin-producing PGPRs act as biocontrol agents against a variety of pathogens [128].

PGPR has been proven in various investigations to be effective in both creating and regulating the amounts of ABA and gibberellic acid in plants. Gibberellins promote primary root elongation and lateral root development. Several PGPR, including Azotobacterspp, Azospirillumspp, Achromobacterxylosoxidans, Gluconobacterdiazotrophicus, Acinetobactercalcoaceticus, Bacillus spp., and Rhizobia spp., have been found to produce gibberellin [129].

The role of ABA under drought stress, for example, is well-known. Under conditions of water deficit, increased ABA levels cause stomata to shut, limiting water loss. This hormone, on the other hand, offers a variety of benefits during lateral root development [129]. Inoculation with Azospirillumbrasilense Sp245 increased ABA content in Arabidopsis, especially when grown under osmotic stress [130].

In addition to their roles in plant RSA, these two hormones are involved in plant defense mechanisms. As a result, PGPR, which produces these hormones, may affect the hormonal balance involved in plant defense, including the jasmonate and salicylic acid pathways [131].

7. PGPR and abiotic stress tolerance

As climate change conditions worsen, extreme environmental conditions that can cause significant annual losses in total crop output are now more prevalent worldwide [132, 133]. Many biotic and abiotic stresses are causing havoc on the sector, resulting in enormous plant productivity losses all around the world. Stress factors comprise nutrient shortages, heavy metal contamination, high wind, extreme temperatures, salinity, drought, illnesses, plant invasions, pests, salt, and soil erosion [69].

As a result of climate change, abiotic stresses, such as drought and high temperatures, have risen in frequency and intensity, resulting in 70% losses in major staple food crops, posing a danger to global food security [134]. Drought and high soil salinity, as well as their downstream impacts, such as osmotic, oxidative, and ionic stress, are regarded as important hindrances to long-term agriculture production [135]. Stressed plants suffer from internal metabolism disruption due to metabolic enzyme inhibition, substrate scarcity, excessive need for different chemicals, or a combo of the following. To endure unfavorable conditions, metabolic reconfiguration is required to comply with the demand for anti-stress compounds, such as suitable solutes, antioxidants, and proteins [136].

Agricultural breeding practices have tried to produce species that are more productive in unfavorable environments for ages. However, crop breeding for abiotic stress resistance has been impeded by a lack of reliable and consistent traits. Tolerance to stress is influenced by a number of genes working together. Furthermore, using agrochemicals to address biotic stresses and nutritional deficits contributes to environmental degradation, has a negative influence on the biogeochemical cycle system, and puts people at increased risk. The potential repercussions of the aforementioned stresses are significant, necessitating the development of robust, cost-effective, and environmentally acceptable methods to mitigate these stresses' harmful effects on plants. As a result, there has been a spike in interest in environmentally friendly and organic agriculture techniques. Plant growth stimulants have been utilized in bio-fertilization, root revitalization, rhizoremediation, disease resistance, and other modes of microbial revival [137].

The efficient approach of PGPR can alleviate stresses that cause serious damage to crop yield, hence, the application of PGPR and/or their byproducts, which can help plants successfully resist extreme environmental circumstances, is one of the most eco-friendly ways [138]. Some PGPR genera, for instance, P. fluorescens, produce the enzymes 1-aminocyclopropane-1-carboxylate (ACC) deaminase and hydroxyacetophenone monooxygenase, which break down the ethylene precursor ACC to a-ketobutyrate and ammonia, thereby protecting plants from abiotic stressors [139]. The most destructive factors that reduce agricultural productivity are salinity and drought [140]. Furthermore, greater ethylene levels in the plant lead to premature fatuity symptoms, including leaf yellowing, abscission, and desiccation/necrosis [141]. PGPR is essential to minimize ethylene concentrations in plants, which in turn reduces stress.

During dry spells, turgor pressure and water potential have a significant impact on plant functionality. Drought stress results in substantial losses in agricultural output and the flow of nutrients, such as sulfates, nitrates, calcium, silica, and magnesium, as well as a reduction in photosynthesis activity [142]. To achieve sustainable agricultural productivity, bacterial colonies in the rhizosphere and endorhizosphere stimulate the plant to withstand drought [143]. PGPR releases osmolytes, which function in tandem with those obtained from plants to keep plants healthy and improve their growth and development, as well as withstand drought-related stress and excessive salt levels in the soil [144]. According to research findings, inoculating plants growing in dry and semi-arid areas with beneficial plant-growth-promoting rhizobacteria (PGPR), which enhances plant abiotic stress tolerance with an osmotic component, could improve drought tolerance and water utilization efficiency. PGPR-induced root development, nutrient uptake efficiency, and systemic tolerance have been proposed as biochemical changes in plants that result in increased abiotic stress tolerance (IST) [78].

8. Plant biotic stress, pesticide use, and PGPR as biopesticides

Rise in global temperature and fluctuations in precipitation as a result of climate change have resulted in unprecedented crop pests and illnesses in various parts of the world [82]. Biotic agents, such as pathogenic bacteria, viruses, fungi, nematodes, protists, weeds, insects, and arachnids, are a prevalent concern in crop production and a long-term danger to sustainable agriculture and ecosystem stability around the world [145]. These species can induce biotic stress in their hosts by interfering with normal metabolism, injuring their plant hosts, reducing plant vigor, limiting plant development, and/or inducing plant mortality. Biotic stress has an impact on co-evolution, ecosystem nutrient cycling, population dynamics, horticulture plant health, and natural habitat ecology [146]. They also result in pre- and post-harvest damage to agricultural crops [147].

According to the FAO, pests are estimated to be responsible for up to 40% of global agricultural production losses each year. Plant diseases cost the world economy more than $220 billion per year while invading insects cost at least $70 billion [148].

Pesticides are chemical compounds that are used to prevent or control pests. However, these are poisonous compounds that pollute soil, watercourses, and plant life. The inappropriate application and overuse of such chemicals have triggered numerous problems (e.g., the emergence of resistance in target organisms, food contamination, and environmental pollution) [149]. Pesticide use causes morphological, physiological, biochemical, and molecular changes in plants that can have a detrimental effect on the plant's development and growth, leaving chemical residues in numerous plant tissues, as well as insect resistance to pesticides [150, 151]. Besides, pesticides cause oxidative stress in plants, hinder physiological and biochemical pathways, cause toxicity, obstruct photosynthesis, and reduce crop yield. The over-generation of reactive oxygen species has a negative effect on non-targeted plants. Reactive oxygen species are highly reactive in nature, causing oxidative damage to lipids, nucleic acids, proteins, carbohydrates, and DNA in plants, as well as disruptions in other biochemical and physiological cell processes [152].

The rising number and intensity of pesticide consumption have presented a significant obstacle to the pests being targeted, leading them to disseminate to dynamic habitats and/or adjust to the changing settings [153]. Resistance is currently the greatest serious impediment to the effective use of pesticides. Many pest species have developed resistance to pesticides as a result of their use around the World [154].

Pesticides' impact on non-target species has been a source of debate and worry around the world for decades. Pesticides' adverse impacts on non-target arthropods have been well documented [155]. Natural insect adversaries, such as parasitoids and predators, are tragically the most vulnerable to insecticides and suffer the most harm [156]. Natural enemies that ordinarily keep small pests in check are sometimes harmed, which can lead to subsequent pest outbreaks.

Not just that, pesticide use may have a negative impact on the earthworm population. Earthworms contribute to the improvement and maintenance of soil structure by producing channels in the soil that allow for aeration and drainage. In agricultural settings, they are regarded as a key indicator of soil quality [157]. Earthworms are harmed by a wide range of agricultural practices, with indiscriminate pesticide usage being one of the most serious [158]. Yasmin and D'Souza [159] found that pesticides have a dose-dependent effect on earthworm reproduction and proliferation.

Moreover, pesticide usage has the potential to destroy biodiversity. Degraded pesticides interface with the soil as well as its inhabitants, affecting microbial diversity, biochemical processes, and enzyme activity [160]. Any change in the activity of soil microorganisms as a result of pesticide application disrupts the ecological environment, resulting in a loss of soil quality. In crops cultivated on soils excessively exposed to chemical pesticides, nutrient loss and disease incidence are widespread [161], which is unfavorable from the perspective of agricultural soil management for food and nutritional security.

Exogenous pesticide residues may also alter the efficacy of beneficial root-colonizing microbes, such as fungi, bacteria, algae, and arbuscular mycorrhiza (AM), in soil by affecting their growth, and metabolic activity, among other things [162].

Furthermore, pesticides are widely distributed when they are transported across long distances by air or water [163]. Several pesticides have a prolonged half-life (up to years) in the environment; for example, the half-life of HCH in water is determined to be 191 days [164], hence posing a threat to aquatic creatures.

The mode of pesticides' action is hazardous not just to the target organisms but also to non-target creatures, such as humans. The physicochemical parameters of the active ingredient are known to influence pesticide diffusion into plant tissue. As a result, pesticides with a systemic effect are absorbed by the roots or leaves and transported throughout the plant, as a result, they pose a major health risk to anyone who consumes them [165]. Pesticides' negative impacts on human health have begun to emerge as a result of their toxicity, longevity in the environment, and tendency to penetrate the food chain. Based on the side effects, chemical pesticides employed in crop protection to limit the damage caused by pathogens and pests in agricultural areas pose significant long-term risks and challenges to life forms. Pesticides can penetrate the human body through immediate exposure to chemicals, contaminated water, or polluted air, as well as through food, particularly fruits and vegetables. Pesticide exposure can cause both acute and chronic disorders. Humans develop chronic sickness after being exposed to sub-lethal levels of pesticides for extended periods of time [166]. They are believed to stimulate cancer [167] and fetal malformations [168], and they are nonbiodegradable [169]. Encountering pesticides with genetic makeup, resulting in DNA damage and chromosomal abnormalities, is one of the primary pathways that lead to chronic disorders, such as cancer [170]. Pesticides can also cause oxidative stress by modifying the amounts of antioxidant enzymes, including glutathione reductase, superoxide dismutase, and catalase, which increase reactive oxygen species (ROS) [171]. Pesticide-induced oxidative stress has been linked to a number of health concerns, including Parkinson's disease and glucose homeostasis disruption [170].

Given the pervasive harmful effects of pesticides on plants, soil, the environment, and human health, an environmentally friendly replacement is required, making PGPR a viable option.

8.1 Biopesticides/biocontrol agents using PGPR

Biocontrol agents are bacteria that suppress the occurrence or severity of plant diseases, whereas antagonists are bacteria that have antagonistic behavior toward a pathogen. PGPR can be used as a biocontrol agent (**Figure 3**) to protect plants from pathogens, such as viruses, bacteria, insects, and fungi [173].

When compared to chemical pesticides, PGPR has unique benefits, including being harmless to mankind and nature, dissolving more quickly in soil, and having a lesser possibility of pathogen resistance development [174]. Because plants, unlike animals, are unable to use avoidance and escape as stress-relieving strategies, their

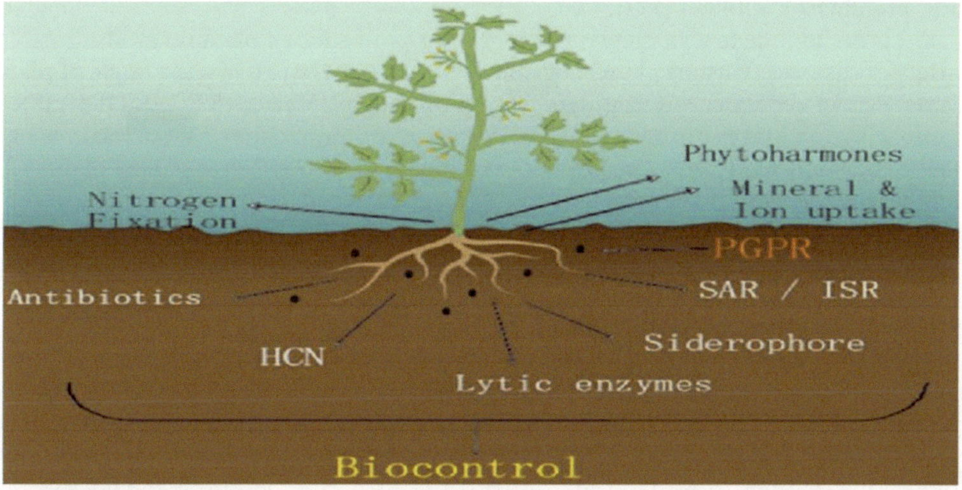

Figure 3.
PGPR as biocontrol agent [172].

existence has been marked by the establishment of extraordinarily favorable partnerships with their more mobile partners, microbes. PGPR and its interactions with plants are economically harnessed [175], and they hold considerable promise for long-term agricultural sustainability. Plants that have been inoculated by immersing their roots or seeds in PGPR cultures overnight have been shown to be extremely resistant to many forms of biotic stress [176].

8.1.1 Antibiotic synthesis

Antibiotic synthesis is one of the most robust and well-studied biocontrol mechanisms of PGPR against phytopathogens during the last two decades [177]. Antibiotics are low-molecular-weight toxins that have the ability to kill or inhibit the growth of other bacteria. The Bacillus genus and Rhizobacteria are the most significant for antibiotic synthesis [178]. Antibacterial and antifungal antibiotics are produced by Bacillus amyloliquefaciens and B. subtilis, including subtilin, bacilysin, and emicobacillin [179].

8.1.2 Induced systemic resistance

Induced systemic resistance (ISR) is a physiological condition of increased defensive capacity triggered by a specific environmental stimulation. Conrath et al. [180] define ISR as "an enhanced defensive ability of plants in response to specific pathogens stimulated by beneficial microorganisms present in the rhizosphere," a scenario wherein the interaction of certain microorganisms with roots results in plant tolerance to pathogenic bacteria, fungi, and viruses. ISR can also be induced by certain environmental cues that cause upregulation of plants' innate defenses in response to the biotic assault, allowing plants to respond faster and stronger to subsequent pathogen attacks [181]. Following the pathogenic invasion, signals are produced, and a defense mechanism is activated via the vascular system. Among the defense mechanisms produced by ISR in plants are cell wall reinforcement [182], production of secondary metabolites, and accumulation of defense-related enzymes, such as chitinases, glucanases, peroxidase, phenylalanine ammonia-lyase, and polyphenol oxidase, lipoxygenase, SOD, CAT, and APX along with some proteinase inhibitors [183].

ISR is not unique to a particular pathogen but can benefit a plant by evading a variety of diseases. Various plants develop systemic resistance to a wide range of plant diseases and a variety of environmental stresses when primed with PGPR [184]. ISR is among the pathways through which PGPR might minimize the onset of various plant diseases by modifying the physical and biochemical attributes of host plants and thereby boosting plant growth [185]. After applying plant-growth-promoting rhizobacteria, diseases of fungal, bacterial, and viral origin, as well as damage caused by insects and nematodes, can be decreased [186].

Non-pathogenic microorganisms promote ISR, which starts in the roots and extends to the shoots [187]. ISR stimulates plant defense mechanisms and shields unexposed regions of plants against future pathogenic attacks by microbes and insects. The signaling of ethylene and jasmonic acid in the plant is involved in induced systemic resistance, and these hormones increase the host plant's defense responses against a range of plant diseases [188]. Lipopolysaccharides (LPS), siderophores, homoserine lactones, 2, 4-diacetylphloroglucinol, cyclic lipopeptides, and volatiles like acetoin and 2, 3-butanediol are only a few of the bacterial components that cause induced systemic resistance [189].

8.1.3 Production of protective enzymes

Plant-growth-promoting rhizobacteria use another mechanism to promote growth—enzymatic activity, producing compounds that inhibit phytopathogenic agents [190]. Rhizobacterial strains that promote plant growth can secrete enzymes, including ACC-deaminase, phosphatases, chitinases, 1,3-glucanase, proteases, dehydrogenases, and lipases, among others [94, 191]. They excrete cell wall hydrolases, which are used to break down cell walls, neutralize infections, assault pathogens, and cause hyperparasitic activity [192]. Plant-growth-promoting rhizobacteria suppress pathogenic fungi, such as *Phytophthora sp, Botrytis cinerea, Fusarium oxysporum, Sclerotium rolfsii*, *Pythium ultimum*, and *Rhizoctonia solani* by the activation of such enzymes [193, 194], hence defending the plant against various biotic and abiotic stresses. Because 1,4-N-acetylglucosamine and chitin make up the majority of fungal cell wall constituents, bacteria that generate 1,3-glucanase and chitinase restrict their proliferation. Inoculation of plants with arbuscular mycorrhiza has also been shown to increase plant growth. *Trichoderma* strains have been employed as biological control agents and plant growth boosters in the past [195].

8.1.4 Production of volatile organic compounds (VOCs)

In recent years, microbial volatile organic compounds (mVOC) have been shown to play an important role in microorganism–plant interactions [196–198]. VOCs are produced by a wide range of soil microorganisms. Bacillus bacteria are the most common microbes that produce antimicrobial MVOCs. Bacterial volatiles have a key function in encouraging plant growth by regulating phytohormone synthesis and metabolism.

They can also promote plant health by acting as antibacterial, nematicidal, oomyceticidal, and antifungal agents, as well as eliciting plant immunity via the salicylic acid (SA) and jasmonic acid (JA) pathways [199]. These molecules have the potential to increase plant growth and development and induce systemic resistance (ISR) against pathogenic organisms, resulting in improved agricultural well-being [200]. Through the SA-signaling pathway, acetoin from the bacteria B. subtilis produces systemic resistance in *A. thaliana* against *P. syringae* [201].

Depending on the species, the quantity and composition of VOCs varies [202]. 2, 3-Butanediol is a volatile organic compound (VOC) generated by a variety of microorganisms that, among other things, can activate plant resistance against pathogens. This mVOC generated by *B. subtilis* and *B. amyloliquefaciens* is capable of generating a systemic resistance in A. thaliana mediated by the ethylene (ET)-signaling pathway against *Erwinia carotovora* subsp. *carotovora* [196]. The same MVOC from *Enterobacter aerogenes* was engaged in the establishment of plant tolerance against *Setosphaeria turcica,* a fungus that causes Northern corn leaf blight [203].

8.1.5 Hydrogen cyanide (HCN) production

The antagonistic activity of PGPR also results in the production of volatile compounds. HCN, a well-studied biocontrol agent, commonly known as prussic acid, is a broad-spectrum volatile secondary metabolite generated by numerous rhizobacteria and is crucial for the biological control of several infectious microorganisms in the soil. Most metalloenzymes are inhibited by their cyanide ion, particularly copper-containing cytochrome c oxidases [204]. HCN-producing Pseudomonas strains are employed in the biological control of tomato bacterial canker [205]. For instance, the inhibition of Macrophomina phaseolina and Meloidogyne javanica caused sunflower charcoal rot and tomato root-knot diseases and has been related to bacterial strains secreting HCN [206]. The inhibitory activity process starts in the mitochondria, where HCN inhibits electron transport, reducing energy supply to the cell and finally causing pathogenic organisms to die.

8.1.6 Aminocyclopropane-1-carboxylate (ACC) deaminase production

Plants generate a lot of "stress ethylene" (ET) after the onset of a disease or stress. Much of the growth inhibition that happens as a result of environmental stress is due to the plant's response to elevated levels of stress ethylene, which aggravates the stressor's response. Likewise, ethylene production inhibitors can considerably reduce the intensity of various environmental stressors. The production of defense enzymes, including 1-aminocyclopropane-1-carboxylate (ACC) deaminase, has also been linked to PGPM's ability to protect against biotic stress [207]. Numerous results suggest that seed inoculation with bacterial endophytes increases plant defense. This is because bacteria produce the enzyme 1-aminocyclopropane-1-carboxylate (ACC), which can cleave ET into ketobutyrate and ammonia, lowering the presence of this enzyme linked to plant stress and physiological impairment [208]. As a result, if ACC deaminase-containing bacteria can reduce plant ethylene levels, treating plants with these organisms may give some defense against the stress inhibitory effects. The synthesis of ACC-deaminase by *Paenibacillus lentimorbus* B-30488 (B-30488) is assumed to be the pathway whereby *P. lentimorbus* B-30488 (B-30488) negates Scelerotium rolfsii in tomato [207]. Hence, the usage of PGPR is appropriate for reducing the environmental stress that crop plants face.

9. Conclusion

To meet the ever-increasing nutritional demand of the rapidly increasing world population, chemical fertilizers must be employed. However, unintended and excessive use has a variety of negative repercussions on the natural environment resulting in soil degradation, global warming, and climate change, necessitating the search for environmentally sound alternatives. PGPR, in this regard, is a realistic choice for agricultural production that does not deplete natural resources. Plants and microbial communities in the soil have evolved a variety of biotic connections, ranging from commensalism to mutualism. Plant-PGPR collaboration is an important aspect of this web of interactions, promoting the growth and health of a variety of plants. PGPR has recently received a lot of attention for its potential to replace agrochemicals for plant growth and yield through a variety of processes, including decomposition of organic matter, recycling of essential elements, formation of soil structure, production of numerous plant growth regulators, fixation of atmospheric nitrogen, degradation of organic pollutants, stimulation of root growth, solubilization of phosphorus, production of siderophore, and solubilization of mineral nutrients, all of which are important for soil and plant health. Furthermore, they are cost-efficient and environmentally sustainable and assure that nutrients from natural sources are always accessible. Besides, bacterial colonies in the rhizosphere have a considerable impact on phytopathogenic microorganism reduction, in addition to boosting plant growth through active processes, hence the use of phytomicrobiome representatives in farming production as long-term disease prevention and nutrient supplement strategy could also help to mitigate the detrimental effects of pesticide use.

As a nutshell, in the face of global climate change, PGPR could be a more environmentally friendly option than chemical fertilizers.

Author details

Tuba Arjumend[1*], Ercüment Osman Sarıhan[2] and Mehmet Uğur Yıldırım[2]

1 Faculty of Agriculture, Department of Plant Protection, Uşak University, Uşak, Turkey

2 Faculty of Agriculture, Department of Field Crops, Usak University, Uşak, Turkey

*Address all correspondence to: tuba.arjumend@usak.edu.tr

References

[1] Roser M. Future population Growth. 2020. Available from: https://OurWorldInData.org; https://ourworldindata.org/future-population-growth [Accessed February 20, 2022]

[2] Zabel F, Delzeit R, Schneider JM, Seppelt R, Mauser W, Václavík T. Global impacts of future cropland expansion and intensification on agricultural markets and biodiversity. Nature Communications. 2019;**10**(1):1-10

[3] Springmann M, Clark M, Mason-D'Croz D, Wiebe K, Bodirsky BL, Lassaletta L, et al. Options for keeping the food system within environmental limits. Nature. 2018;**562**:519-525. DOI: 10.1038/s41586-41018-40594-41580

[4] Brentrup F. The impact of mineral fertilizers on the carbon footprint of crop production. In: Conference Paper. International Plant Nutrition Colloquium XVI. Davis/US: University of California; 2009

[5] Randive K, Raut T, Jawadand S. An overview of the global fertilizer trends and India's position in 2020. Mineral Economics. 2021;**34**(3):371-384

[6] Williams J. The Worldwide Fertilizer Market to 2024. 2019. Avaiable from: https://www.World fertilizer.com/environment/21032019/the-worldwidefertilizer-market-to-2024/2019 [Accessed February 25, 2022]

[7] IFA. World Fetilizer Consumption Statistics, Annual. 2021

[8] Available from: https://www.statista.com/statistics/438967/fertilizer-consumption-globally-by-nutrient/ [Accessed February 20. 2022]

[9] TUIK. Turkish Statistical Institute. Available from: https://data.tuik.gov.tr/Search/Search?text=gübre [Accessed November 10, 2021]

[10] Taban S. Gübre Kullanımının Öyküsü. Mart-Haziran: Tarım ve Mühendislik Dergisi; 2018

[11] Shoji S, Delgado J, Mosier A, Miura Y. Use of controlled release fertilizers and nitrification inhibitors to increase nitrogen use efficiency and to conserve air and water quality. Communications in Soil Science and Plant Analysis. 2001;**32**(7-8):1051-1070

[12] Riegel M. Chemical fertilizers is harmful not only to the environment, but also to surrounding water sources: Marine pollution. Journal of Pollution Effects & Control. 2021;**9**:285. DOI: 10.35248/2375-4397.21.9.285

[13] Kotschi J. Adverse Impacts of Mineral Fertilizers in Tropical Agriculture. A soiled reputation. A study Commission. Germany: Heinrich Böll Foundation, WWF; 2015

[14] Şahin G. The fertilizer use state in turkey and problems experienced about fertilizing. Turkish Journal Agricultural Economics. 2016;**22**(1):19-32

[15] Bisht N, Chauhan PS. Excessive and disproportionate use of chemicals cause soil contamination and nutritional stress. In: Soil Contamination-Threats and Sustainable Solutions. London: IntechOpen; 2020. DOI: 10.5772/intechopen.94593

[16] Chandini KR, Kumar R, Prakash O. The impact of chemical fertilizers on our environment and ecosystem. Research Trends in Environmental Sciences. 2019, 2019;**2**:69-86

[17] Savci S. Investigation of effect of chemical fertilizers on environment.

APCBEE Procedia. 2012;**1**:287-292. DOI: 10.1016/j.apcbee.2012.03.047

[18] Hao T, Zhu Q, Zeng M, Shen J, Shi X, Liu X, et al. Impacts of nitrogen fertilizer type and application rate on soil acidification rate under a wheat-maize double cropping system. Journal of Environmental Management. 2020;**270**:110888

[19] Han J, Shi J, Zeng L, Xu J, Wu L. Effects of nitrogen fertilization on the acidity and salinity of greenhouse soils. Environmental Science and Pollution Research. 2015;**22**(4):2976-2986

[20] Li XH, Han XZ, Li HB, Song C, Yan J, Liang Y. Soil chemical and biological properties affected by 21-year application of composted manure with chemical fertilizers in a Chinese Mollisol. Canadian Journal of Soil Science. 2012;**92**(3):419-428

[21] Al-Busaidi W, Janke R, Menezes-Blackburn D, Khan MM. Impact of long-term agricultural farming on soil and water chemical properties: A case study from Al-Batinah regions (Oman). Journal of the Saudi Society of Agricultural Sciences. 2021. DOI: 10.1016/j.jssas.2021.11.002

[22] Sönmez İ, Sönmez S. Tuzluluk ve Gübreleme Arasındaki İlişkiler. Tarımın Sesi Dergisi. 2007;**16**:13-16

[23] Köleli N, Kantar Ç. Fosforlu Gübrelerde Ağır Metal Tehlikesi. Ekoloji Dergisi. 2006;**9**:1-5

[24] Masindi V, Muedi K. Environmental Contamination by Heavy Metals, Heavy Metals. London: IntechOpen; 2018

[25] Lei D, Liang D, Yu Y, Chen W, Song J. Human health risk assessment of heavy metals in the irrigated area of Jinghui, Shaanxi, China, in terms of wheat flour consumption environ. Environmental Monitoring and Assessment. 2015;**187**:647

[26] Li H, Liu J, Li G, Shen J, Bergström L, Zhang F. Past, present, and future use of phosphorus in Chinese agriculture and its influence on phosphorus losses. Ambio. 2015;**44**:274-285

[27] Loganathan P, Hedley MJ, Grace ND. Pasture soils contaminated with fertilizer-derived cadmium and fluorine: Livestock effects. Reviews of Environmental Contamination and Toxicology. 2008;**192**:29-66

[28] Chen XX, Liu YM, Zhao QY, Cao WQ, Chen XP, Zou CQ. Health risk assessment associated with heavy metal accumulation in wheat after long-term phosphorus fertilizer application. Environmental Pollution. 2020;**262**:114348

[29] Jiao W, Chen W, Chang AC, Page AL. Environmental risks of trace elements associated with long-term phosphate fertilizers applications: A review. Environmental Pollution. 2012, 2012;**168**:44-53

[30] Chary N, Kamala CT, Samuel Suman Raj D. Assessing risk of heavy metals from consuming food grown on sewage irrigated soils and food chain transfer. Ecotoxicology and Environmental Safety. 2008;**69**:513-524

[31] Huang SW, Jin JY. Status of heavy metals in agricultural soils as affected by different patterns of land use. Environmental Monitoring and Assessment. 2008;**139**(1-3):317-327. DOI: 10.1007/s10661-007-9838-4

[32] Atafar Z, Mesdaghinia A, Nouri J, Homaee M, Yunesian M, Ahmadimoghaddam M, et al. Effect of fertilizer application on soil heavy

metal concentration. Environmental Monitoring and Assessment. 2010;**160**(1):83-89

[33] Wei B, Yu J, Cao Z, Meng M, Yang L, Chen Q. The availability and accumulation of heavy metals in greenhouse soils associated with intensive fertilizer application. International Journal of Environmental Research and Public Health. 2020;**17**(15):5359

[34] Küçük Ahmetler O. The effects of salinity on yield and quality of ornamental plants and cut flowers. In: ISHS Acta Horticulturae 573: International Symposium on Techniques to Control Salination for Horticultural Productivity. Leuven, Belgium: ISHS Acta Horticulturae; 2000. pp. 407-414. DOI: 10.17660/ActaHortic.2002.573.49

[35] Özbek H, Kaya Z, Gök M, Kaptan H. Toprak Bilimi. Adana: Çukurova Üniversitesi Ziraat Fakültesi Ders Kitapları Yayınları; 1999. pp. 574-575

[36] Welle PD, Mauter MS. High-resolution model for estimating the economic and policy implications of agricultural soil salinization in California. Environmental Research Letters. 2017;**12**:094010. DOI:10.1088/1748-9326/aa848e

[37] Topbaş MT, Brohi AR, Karaman MR. ÇevreKirliliği. Ankara: T.C. ÇevreBakanlığıYayınları; 1998

[38] Weisskopf P, Reiser R, Rek J, Oberholzer HR. Effect on different compaction impacts and varying subsequent management practices on soil structure, air regime and microbiological parameters. Soil & Tillage Research. 2010;**111**:65-74

[39] Pahalvi HN, Rafiya L, Rashid S, Nisar B, Kamili AN. Chemical fertilizers and their impact on soil health. In: Microbiota and Biofertilizers. 2021; 2. Cham: Springer; 2021. pp. 1-20

[40] Blanco CH, Gantzer CH, Anderson SH, Alberts EE, Ghidey F. Saturated hydraulic conductivity and its impact on simulated runoff for claypan soils. Soil Science Society of America Journal. 2002;**66**:1596-1602

[41] Batey T. Soil compaction and soil management -a review. Soil Use and Management. 2009;**25**(4):335-345

[42] Rannik K. 2009. Soil compaction effects on soil bulk density and penetration resistance and growth of spring barley (Hordeum vulgare L.). Journal of Acta Agriculturae Scandinavica Plant Soil Science. 2009;**59**(3):265-272

[43] Barzegar AR, Nadian H, Heidari F, Herbert SJ, Hashemi AM. Interaction of soil compaction, phosphorus and zinc on clover growth and accumulation of phosphorus. Soil & Tillage Research. 2006;**87**:155-162

[44] Massah J, Azadegan B. Effect of chemical fertilizers on soil compaction and degradation. Agricultural Mechanization In Asia, Africa, and Latin America. 2016;**47**(1):44-50

[45] Saber MSM. CleanBiotechnology forsustainablefarming. Engineering in Life Sciences. 2001;**1**:217-223

[46] Bøckman OC. Fertilizers and biological nitrogen fixation as sources of plant nutrients: Perspectives for future agriculture. Plant Soil. 1997;**194**:11-14

[47] Sun M, Xiao T, Ning Z, Xiao E, Sun W. Microbial community analysis in rice paddy soils irrigated by acid mine drainage contaminated water. Applied Microbiology and Biotechnology. 2015;**99**:2911-2922. DOI: 10.1007/s00253-014-6194-5

[48] Singh JS, Gupta VK. Soil microbial biomass: A key soil driver in management of ecosystem functioning. Science of the Total Environment. 2018;**634**:497-500. DOI: 10.1016/j.scitotenv.2018.03.373

[49] Yıldız NA. Tarımsal faaliyetlerin çevre kirliliği üzerine etkileri. Availabe from: https://atauni.edu.tr/yuklemeler/ab0101aee211b1d48876061170edfe80.pdf [Accessed February 25, 2022]

[50] Sönmez İ, Kaplan M, Sönmez S. Kimyasal gübrelerin çevre kirliliği üzerine etkileri ve çözüm önerileri. Batı Akdeniz Tarımsal Araştırma Enstitüsü Derim Dergisi. 2008;**25**:24-34 ISSN 1300-3496

[51] Korkmaz K. Tarım Girdi Sisteminde Azot ve Azot Kirliliği. 2007. Available from:http://www.ziraat.ktu.edu.tr/tarim_girdi.htm

[52] Savci S. An agricultural pollutant: chemical fertilizer. International Journal of Environmental Science and Development. 2012;**3**(1):73

[53] Taşkaya B. Tarımve Çevre. Tarımsal Ekonomi Araştırma Enstitüsü. 2004;**5**: 1-8. ISSN 1303-8346

[54] Rütting T, Aronsson H, Delin S. Efficient use of nitrogen in agriculture. Nutrient Cycling in Agroecosystems. 2018;**110**:1-5

[55] Sharma A, Chetani R. A review on the effect of organic and chemical fertilizers on plants. International Journal for Research in Applied Science & Engineering Technology (IJRASET). 2017;**677**

[56] Chen J-H. The combined use of chemical and organic fertilizers and/or biofertilizer for crop growth and soil fertility. In: International Workshop on Sustained Management of the Soil-rhizosphere System for Efficient Crop Production and Fertilizer Use. Food and Fertilizer Technology Center: Taipei, Taiwan; 2006. pp. 16-20

[57] Çakmakçı R. BitkiGelişiminiTeşvik Eden Rizobakterilerin Tarımda Kullanımı (Use of Plant Growth Promoting Rhizobacteria in Agriculture). Atatürk Üniversitesi Ziraat Fakültesi Dergisi. 2005;**36**(1):97-107

[58] Lobell DB, Schlenker W, Costa-Roberts J. Climate trends and global crop production since 1980. Science. 2011;**333**:616-620. DOI: 10.1126/science.1204531

[59] Lipper L, Thornton P, Campbell BM, Baedeker T, Braimoh A, Bwalya M, et al. Climate-smart agriculture for food security. Nature Climate Change. 2014;**4**:1068-1072. DOI: 10.1038/nclimate2437

[60] Drost S, Rutgers M, Wouterse M, De Boer W, Bodelier PL. Decomposition of mixtures of cover crop residues increases microbial functional diversity. Geoderma. 2020;**361**:114060. DOI: 10.1016/j.geoderma.2019.114060

[61] Leach JE, Triplett LR, Argueso CT, Trivedi P. Communication in the phytobiome. Cell. 2017;**169**:587-596

[62] Hiltner L. Uber neuereerfahrungen und probleme auf demgebiet der bodenbakteriologie und unterbesonderer berucksichtigungdetgrundungung und branche. Arbeiten der Deutschen Landwirtschafts - Gesellschaf. 1904;**98**:59-78

[63] Beattie GA. Microbiomes: Curating communities from plants. Nature. 2015;**528**:340-341. DOI: 10.1038/nature16319

[64] Bender SF, Wagg C, Van Der Heijden MG. An underground

revolution: Biodiversity and soil ecological engineering for agricultural sustainability. Trends in Ecology & Evolution. 2016;**31**:440-452. DOI: 10.1016/j.tree.2016.02.016

[65] Boroujeni YK, Boroujeni VN, Rastegari AA, Yadav N, Yadav AN. Soil microbes with multifarious plant growth promoting attributes for enhanced production of food crops. In: Soil Microbiomes for Sustainable Agriculture. Champions: Springer; 2021. pp. 55-83

[66] Egamberdieva D, Hashem A, Alqarawi AA. Microbial phytohormones have a key role in mitigating the salt-induced damages in plants. In: Maheshwari DK editor Bacterial Metabolites in Sustainable Agroecosystem.Cham: Springer International Publishing. 2015;283-296. DOI: 10.1007/978-3-319-24654-3_10

[67] Kumar A, Dubey A. Rhizosphere microbiome: Engineering bacterial competitiveness for enhancing crop production. Journal of Advanced Research. 2020;**24**:337-352. DOI: 10.1016/j.jare.2020.04.014

[68] Vessey JK. Plant growth promoting rhizobacteria as biofertilizers. Plant and Soil. 2003;**255**:571-586. DOI: 10.1023/A:1026037216893

[69] Turan M, Arjumend T, Argın S, Yıldırım E, Katırcıoğlu H, Gürkan B, et al. Plant root enhancement by plant growth promoting rhizobacteria. In: Plant Roots. Vol. 2021. London: IntechOpen; 2021. DOI: 10.5772/intechopen.99890

[70] Bibi N, Hamayun M, Khan SA, qbal A, Islam B, Shah F, et al. Anthracene biodegradation capacity of newly isolated rhizospheric bacteria Bacillus cereus S13. PLoS One. 2018;**13**:e0201620. DOI: 10.1371/journal.pone.0201620

[71] Pii Y, Mimmo T, Tomasi N, Terzano R, Cesco S, Crecchio C. Microbial interactions in the rhizosphere: Beneficial influences of plant growth-promoting rhizobacteria on nutrient acquisition process. A review. Biology and Fertility of Soils. 2015;**51**:403-415. DOI: 10.1007/s00374-015-0996-1

[72] Yin C, Casa Vargas JM, Schlatter DC, Hagerty CH, Hulbert SH, Paulitz TC. Rhizosphere community selection reveals bacteria associated with reduced root disease. Microbiome. 2021;**9**(1):1-18

[73] Akhtar N, Qureshi MA, Iqbal A, Ahmad MJ, Khan KH. Influence of azotobacter and IAA on symbiotic performance of rhizobium and yield parameters of lentil. Journal of Agricultural Research. 2012;**50**:361-372

[74] Sivasakhti S, Usharani G, Saranraj P. Biocontrol potentiality of plant growth promoting bacteria (PGPR)- Pseudomonas fluorescens and Bacillus subtilis: A review. African Journal of Agricultural Research. 2014;**9**:1265-1277

[75] Singh AK, Kumar A, Singh PK. PGPR Amelioration in Sustainable Agriculture: Food Security and Environmental Management. Sawston, United Kingdom: Woodhead Publishing; 2018

[76] Maheshwari DK, Dubey RC, Aeron A, Kumar B, Kumar S, et al. Integrated approach for disease management and growth enhancement of Sesamumindicum L. utilizing Azotobacterchroococcum TRA2 and chemical fertilizer. World Journal of Microbiology and Biotechnology. 2012;**28**:3015-3024

[77] Antar M, Lyu D, Nazari M, Shah A, Zhou X, Smith DL. Biomass for a sustainable bioeconomy: An overview of world biomass production and utilization. Renewable and Sustainable

Energy Reviews. 2021;**139**:110691. DOI: 10.1016/j.rser.2020.110691

[78] Turan M, Ekinci M, Yıldırım E, Güneş A, Karagöz K, Kotan R, et al. Plant growth-promoting rhizobacteria improved growth, nutrient, and hormone content of cabbage (Brassica oleracea) seedlings. Turkish Journal of Agriculture and Forestry. 2014;**38**:327-333. DOI: 10.3906/tar-1308-62

[79] Figueiredo MDVB, Bonifacio A, Rodrigues AC, Araujo FFD. Plant growth-promoting rhizobacteria: Key mechanisms of action. In: Microbial-Mediated Induced Systemic Resistance in Plants. 2016. Singapore: Springer; 2016. pp. 23-37

[80] Alori ET, Dare MO, Babalola OO. Microbial inoculants for soil quality and plant health. In: Lichtfouse E, editor. Sustainable Agriculture Reviews. Cham: Springer; 2017. pp. 281-307. DOI: 10.1007/978-3-319-48006-0_9

[81] Pep C, Tian H, Patra P, Thompson R. Nitrogen Fertilizers Are Incredibly Efficient, But They Make Climate Change a Lot Worse. Melbourne, Australia: The Conversation; 2019. Available from: https://theconversation.com/ nitrogen-fertilizers-are-incredibly-efficient-but-they-make-climate-change a-lot-worse-127103: Accessed February 20, 2022

[82] Naamala J, Smith DL. Relevance of plant growth promoting microorganisms and their derived compounds, in the face of climate change. Agronomy. 2020;**10**:1179. DOI: 10.3390/agronomy10081179

[83] Hillel D. Soil biodiversity. In: Soil in the Environment, edHillelD. San Diego, CA: Academic Press; 2008. pp. 163-174. DOI: 10.1016/B978-0-12-348536-6.50017-4

[84] Turan M, Kitir N, Alkaya Ü, Günes A, Tüfenkçi S, Yildirim E, et al. Making Soil More Accessible to Plants: The Case of Plant Growth Promoting Rhizobacteria. Vol. 5. London: IntechOpen; 2016. pp. 61-69. DOI: 10.5772/64826

[85] Oberson A, Frossard E, Bühlmann C, Mayer J, Mäder P, Lüscher A. Nitrogen fixation and transfer in grass clover leys under organic and conventional cropping systems. Plant and Soil. 2013;**371**:237-255. DOI: 10.1007/ s11104-013-1666-4

[86] Naamala J, Jaiswal SK, Dakora FD. Microsymbiont diversity and phylogeny of native bradyrhizobia associated with soybean (Glycine max L. Merr.) nodulation in South African soils. Systematic and Applied Microbiology. 2016;**39**:336-344. DOI: 10.1016/j.syapm.2016.05.009

[87] Adesemoye AO, Kloepper JW. Plant–microbes interactions in enhanced fertilizer- use efficiency. Applied Microbiology and Biotechnology. 2009;**85**:1-12. DOI: 10.1007/s00253-009-2196-0

[88] Boddey RM, Urquiaga S, Alves BJR, Reis V. Endophytic nitrogen fixation in sugarcane: Present knowledge and future applications. Plant and Soil. 2003;**252**:139-149

[89] Tak HI, Ahmad F, Babalola O, Inam A. Growth, photosynthesis and yield of chickpea as influenced by urban wastewater and different levels of phosphorus. International Journal of Plant Research. 2012;**2**:6-13. DOI: 10.5923/j. plant.20120202.02

[90] Khan MS, Zaidi A, Ahemad M, Oves M, Wani PA. Plant growth promotion by phosphate solubilizing fungi - current perspective. Archives of Agronomy and Soil Science. 2010;**56**:73-98

[91] Pandey P, Maheshwari DK. Two sp. microbial consortium for growth

promotion of CajanusCajan. Current Science. 2007;**92**:1137-1142

[92] Ramaekers L, Remans R, Rao IM, Blair MW, Vanderleyden J. Strategies for improving phosphorus acquisition efficiency of crop plants. Field Crops Research. 2010;**117**(2-3):169-176

[93] Shah A, Smith DL. Flavonoids in agriculture: Chemistry and roles in, biotic and abiotic stress responses, and microbial associations. Agronomy. 2020;**10**:1209. DOI: 10.3390/agronomy10081209

[94] Rifat H, Safdar A, Ummay A, Rabia K, Iftikhar A. Soil beneficial bacteria and their role in plant growth promotion: A review. Annales de Microbiologie. 2010;**60**:579-598

[95] Miller SH, Browne P, Prigent-Combaret C, Combes-Meynet E, Morrissey JP, O'Gara F. Biochemical and genomic comparison of inorganic phosphate solubilization in pseudomonas species. Environmental Microbiology Reports. 2010;**2**(3):403-411

[96] Sundara B, Natarajan V, Hari K. Influence of phosphorus solubilizing bacteria on the changes in soil available phosphorus and sugarcane and sugar yields. Field Crops Research. 2002;**77**:43-49. DOI: 10.1016/S0378-4290(02)00048-5

[97] Khan N, Bano A, Ali S, Babar MA. Crosstalk amongst phytohormones from planta and PGPR under biotic and abiotic stresses. Plant Growth Regulation. 2020;**90**:189-203. DOI: 10.1007/s10725-020-00571-x

[98] Parmar P, Sindhu SS. Potassium solubilization by rhizosphere bacteria: Influence of nutritional and environmental conditions. Journal of Microbiology Research. 2013;**3**:25-31

[99] Han HS, Lee KD. Effect of co-inoculation with phosphate and potassium solubilizing bacteria on mineral uptake and growth of pepper and cucumber. Plant, Soil and Environment. 2006;**52**:130-136

[100] Liu D, Lian B, Dong H. Isolation of Paenibacillus sp. and assessment of its potential for enhancing mineral weathering. Journal of Geomicrobiology. 2012;**29**:413-421

[101] Setiawati TC, Mutmainnah L. Solubilization of potassium containing minerals by microorganisms from sugarcane rhizosphere. Agricultural Science Procedia. 2016;**9**:108-117

[102] Höfte M. Classes of microbial siderophores. In: Barton LL, Hemming BC, editors. Iron Chelation in Plants and Soil Microorganisms. New York: Academic Press; 1992. pp. 3-26 ISBN: 9780323147835

[103] Kloepper JW, Leong J, Teintze M, Schroth MN. Pseudomonas siderophores: A mechanism explaining disease suppressive soils. Current Microbiology. 1980;**4**:317-320

[104] Ma Y, Prasad MNV, Rajkumar M, Freitas H. Plant growth promoting rhizobacteria and endophytes accelerate phytoremediation of metalliferous soils. Biotechnology Advances. 2011;**29**:248-258

[105] Chaiharn M, Chunhaleuchanon S, Lumyong S. Screening siderophores producing bacteria as potential biological control agents for fungal rice pathogens in Thailand. World Journal of Microbiology and Biotechnology. 2009;**25**:1919-1928. DOI: 10.1007/s11274-009-0090-7

[106] Beneduzi A, Ambrosini A, Passaglia LMP. Plant growth-promoting

rhizobacteria: their potential as antagonists and biocontrol agents. Genetics and Molecular Biology. 2012;**35**(4):1044-1051

[107] Gururani MA, Upadhyaya CP, Baskar V, Venkatesh J, Nookaraju A, Park SW. Plant growth-promoting rhizobacteria enhance abiotic stress tolerance in solanumtuberosum through inducing changes in the expression of ROS-scavenging enzymes and improved photosynthetic performance. Journal of Plant Growth Regulation. 2013;**32**:245-258. DOI: 10.1007/s00344-012-9292-6

[108] Qurashi AW, Sabri AN. Bacterial exopolysaccharide and biofilm formation stimulate chickpea growth and soil aggregation under salt stress. Brazilian Journal of Microbiology. 2012;**43**:1183-1191

[109] Tewari S, Arora NK. Multifunctional exopolysaccharides from Pseudomonas aeruginosa PF23 involved in plant growth stimulation, biocontrol and stress amelioration in sunflowers under saline conditions. Current Microbiology. 2014;**69**:484-494

[110] Parada M, Vinardell J, Ollero F, Hidalgo A, Gutiérrez R, et al. Sinorhizobiumfredii HH103 mutants affected in capsular polysaccharide (KPS) are impaired for nodulation with soybean and Cajanuscajan. Molecular Plant-Microbe Interactions. 2006;**19**:43-52

[111] Arora NK, Tewari S, Singh R. Multifaceted plant-associated microbes and their mechanisms diminish the concept of direct and indirect PGPRs. In: Plant Microbe Symbiosis: Fundamentals and Advances. Berlin/Heidelberg, Germany: Springer; 2013. pp. 411-449. DOI: 10.1007/978-81-322-1287-4_16

[112] Damam M, Kaloori K, Gaddam B, Kausar R. Plant growth promoting substances (phytohormones) produced by rhizobacterial strains isolated from the rhizosphere of medicinal plants. International Journal of Pharmaceutical Sciences Review and Research. 2016;**37**(1):130-136

[113] Sureshbabu K, Amaresan N, Kumar K. Amazing multiple function properties of plant growth promoting rhizobacteria in the rhizosphere soil. International Journal of Current Microbiology and Applied Sciences. 2016;**5**(2):661-683

[114] Hu Y, Jiang Y, Han X, Wang H, Pan J, Yu D. (2017). Jasmonate regulates leaf senescence and tolerance to cold stress: Crosstalk with other phytohormones. Journal of Experimental Botany. 2017;**68**(6):1361-1369

[115] Khan AL, Waqas M, Kang SM, Al-Harrasi A, Hussain J, Al-Rawahi A, et al. Bacterial endophyte Sphingomonassp LK11 produces gibberellins and IAA and promotes tomato plant growth. Journal of Microbiology. 2014;**52**:689-695. DOI: 10.1007/s12275-014-4002-7

[116] Backer R, Rokem JS, Ilangumaran G, Lamont J, Praslickova D, Ricci E, et al. Plant growth-promoting rhizobacteria: Context, mechanisms of action, and roadmap to commercialization of biostimulants for sustainable agriculture. Frontiers in Plant Science. 2018;**9**:1473. DOI: 10.3389/fpls.2018.01473

[117] Tanimoto E. Regulation and root growth by plant hormones-roles for auxins and gibberellins. Critical Reviews in Plant Sciences. 2005;**24**:249-265

[118] Asgher M, Khan MIR, Anjum NA, Khan NA. Minimizing toxicity of cadmium in plants-role of plant growth regulators. Protoplasma. 2015;**252**:399-413. DOI: 10.1007/s00709-014-0710-4

[119] Egamberdieva D, Kucharova Z. Selection for root colonizing bacteria stimulating wheat growth in saline soils. Biology and Fertility of Soils. 2009;**45**:561-573. DOI: 10.1007/s00374-009-0366-y

[120] Kumar A, Patel JS, Meena VS, Srivastava R. Recent advances of PGPR based approaches for stress tolerance in plants for sustainable agriculture. Biocatalysis and Agricultural Biotechnology. 2019;**20**:101271. DOI: 10.1016/j.bcab.2019.101271

[121] Saleem M, Asghar HN, Khan MY, Zahir ZA. Gibberellic acid in combination with pressmud enhances the growth of sunflower and stabilizes chromium (VI)-contaminated soil. Environmental Science and Pollution Research. 2015;**22**(14):10610-10617

[122] Spaepen S, Vanderleyden J. Auxin and plant-microbe interactions. Cold Spring Harbor Perspectives in Biology. 2011;**3**:a001438

[123] Jha CK, Saraf M. Plant growth promoting rhizobacteria (PGPR): A review. E3 Journal of Agricultural Research and Development. 2015;**5**:108-119

[124] Akhtar SS, Mekureyaw MF, Pandey C, Roitsch T. Role of cytokinins for interactions of plants with microbial pathogens and pest insects. Frontiers in Plant Science. 2020;**10**:1777. DOI: 10.3389/fpls.2019.01777

[125] Aloni R, Aloni E, Langhans M. Role of cytokinin and auxin in shaping root architecture: Regulating vascular differentiation, lateral root initiation, root apical dominance and root gravitropism. Annals of Botany. 2006;**97**:883-893

[126] Riefler M, Novak O, Strnad M, Schmulling T. Arabidopsis cytokinin receptor mutants reveal functions in shoot growth, leaf senescence, seed size, germination, root development, and cytokinin metabolism. The Plant Cell. 2006;**18**:40-54

[127] Kang SM, Khan AL, Waqas M, You YH, Hamayun M, Joo GJ, et al. Gibberellin-producing Serratianematodiphila PEJ1011 ameliorates low temperature stress in Capsicum annuum L. European Journal of Soil Biology. 2015;**68**:85-93. DOI: 10.1016/j.ejsobi.2015.02.005

[128] Maheshwari DK, Dheeman S, Agarwal M. Phytohormone-producing PGPR for sustainable agriculture. In: Maheshwari DK, editor. Bacterial Metabolites in Sustainable Agroecosystem. Berlin/Heidelberg, Germany: Springer International; 2015. pp. 159-182. DOI: 10.1007/978-3-319-24654-3_7

[129] Dodd I, Zinovkina N, Safronova V, Belimov A. Rhizobacterial mediation of plant hormone status. The Annals of Applied Biology. 2010;**157**:361-379

[130] Cohen AC, Bottini R, Piccoli PN. AzospirillumbrasilenseSp 245 produces ABA in chemically-defined culture medium and increases ABA content in arabidopsis plants. Plant Growth Regulation. 2008;**54**(2):97-103

[131] Pieterse CM, Leon-Reyes A, Van der Ent S, Van Wees SC. (2009). Networking by small-molecule hormones in plant immunity. Nature Chemical Biology. 2009;**5**(5):308-316

[132] Shrivastava P, Kumar R. Soil salinity: A serious environmental issue and plant growth promoting bacteria as one of the tools for its alleviation. Saudi Journal of Biological Sciences. 2015;**22**:123-131. DOI: 10.1016/j.sjbs.2014.12.001

[133] Mishra J, Singh R, Arora NK. Alleviation of heavy metal stress in plants

and remediation of soil by rhizosphere microorganisms. Frontiers in Microbiology. 2017;**8**:1706. DOI: 10.3389/fmicb.2017.01706

[134] Lobell DB, Field CB. Global scale climate–crop yield relationships and the impacts of recent warming. Environmental Research Letters. 2007;**2**:014002

[135] Kaushal M, Wani SP. Rhizobacterial-plant interactions: Strategies ensuring plant growth promotion under drought and salinity stress. Agriculture, Ecosystems & Environment. 2016;**231**:68-78

[136] Obata T, Fernie AR. The use of metabolomics to dissect plant responses to abiotic stresses. Cellular and Molecular Life Sciences. 2012;**69**(19):3225-3243

[137] Turan M, Argin S, Bolouri P, Arjumend T, Ersoy N, Yıldırım E, et al. Bioremediation of contaminated soil with plant growth rhizobium bacteria. In: Assisted Phytoremediation. Vol. 2022. Amsterdam, Netherlands: Elsevier; 2022. pp. 265-284

[138] Asghari B, Khademian R, Sedaghati B. Plant growth promoting rhizobacteria (PGPR) confer drought resistance and stimulate biosynthesis of secondary metabolites in pennyroyal (Mentha pulegium L.) under water shortage condition. Scientia Horticulturae. 2020;**263**:109132. DOI: 10.1016/j.scienta.2019.109132

[139] Kumar A, Maleva M, Bruno LB, Rajkumar M. Synergistic effect of ACC deaminase producing Pseudomonas sp. TR15a and siderophore producing Bacillus aerophilus TR15c for enhanced growth and copper accumulation in Helianthus annuus L. Chemosphere. 2021;**276**:130038. DOI: 10.1016/j.chemosphere.2021.130038

[140] Hasanuzzaman M, Nahar K, Fujita M. Plant response to salt stress and role of exogenous protectants to mitigate salt-induced damages. In: Ecophysiology and Responses of Plants Under Salt Stress. New York, NY: Springer; 2013. pp. 25-87

[141] Gamalero E, Glick BR. Bacterial modulation of plant ethylene levels. Plant Physiology. 2015;**169**(1):13-22

[142] Rahdari P, Hosseini SM, Tavakoli S. The studying effect of drought stress on germination, proline, sugar, lipid, protein and chlorophyll content in purslane (Portulacaoleracea L.) leaves. Journal of Medicinal Plants Research. 2012;**6**(9):1539-1547

[143] Grover M, Ali SZ, Sandhya V, Rasul A, Venkateswarlu B. Role of microorganisms in adaptation of agriculture crops to abiotic stresses. World Journal of Microbiology and Biotechnology. 2011;**27**(5):1231-1240

[144] Vardharajula S, Zulfikar AS, Grover M, Reddy G, Bandi V. Drought-tolerant plant growth promoting bacillus spp: Effect on growth, osmolytes, and antioxidant status of maize under drought stress. Journal of Plant Interactions. 2011;**6**:1-14. DOI: 10.1080/17429145.2010.535178

[145] Haggag WM, Abouziena HF, Abd-El-Kreem F, El Habbasha S. Agriculture biotechnology for management of multiple biotic and abiotic environmental stress in crops. Journal of Chemical and Pharmaceutical Research. 2015;7(10):882-889

[146] Gusain YS, Singh US, Sharma AK. Bacterial mediated amelioration of drought stress in drought tolerant and susceptible cultivars of rice (Oryza sativa L.). African Journal of Biotechnology. 2015;**14**(9):764-773

[147] Singla J, Krattinger SG. Biotic stress resistance genes in wheat. In: Wrigley CW, Faubion J, Corke H, Seetharaman K, editors. Encyclopedia of Food Grains. Oxford: Elsevier; 2016. pp. 388-392

[148] FAO. Available from: https://www.fao.org/news/story/en/item/1402920/icode/ [Accessed February 26, 2022]

[149] Lei YH, Zhang CF, Wang Z, Liu S, Li XS. Residues and risk assessment of environmental hormone pesticide triazole fungicides in soil. Agrochemicals. 2019;**58**:660-663

[150] Sharma A, Kumar V, Kumar R, Shahzad B, Thukral AK, Bhardwaj R. Brassinosteroid-mediated pesticide detoxification in plants: A mini-review. Cogent Food & Agriculture. 2018;**4**(1):1436212

[151] Shahzad B, Tanveer M, Che Z, Rehman A, Cheema SA, Sharma A, et al. Role of 24-epibrassinolide (EBL) in mediating heavy metal and pesticide induced oxidative stress in plants: A review. Ecotoxicology and Environmental Safety. 2018;**147**:935-944

[152] Sharma A, Gupta P, Prabhakar PK. Endogenous repair system of oxidative damage of DNA. Current Chemical Biology. 2019;**13**(2):110-119

[153] Cothran RD, Brown JM, Relyea RA. Proximity to agriculture is correlated with pesticide tolerance: Evidence for the evolution of amphibian resistance to modern pesticides. Evolutionary Applications. 2013;**6**(5):832-841

[154] Tabashnik BE, Van Rensburg JBJ, Carrière Y. Field-evolved insect resistance to Bt crops: Definition, theory, and data. Journal of Economic Entomology. 2009;**102**(6):2011-2025

[155] Ware GW. Effects of pesticides on nontarget organisms. Residue Reviews. 1980:173-201

[156] Vickerman GP. Farm scale evaluation of the long-term effects of different pesticide regimes on the arthropod fauna of winter wheat. In: Fields Methods for the Study of Environmental Effects of Pesticides. Symposium. Farnham, United Kingdom: British Crop Protection Council; 1988. pp. 127-113

[157] Paoletti MG. The role of earthworms for assessment of sustainability and as bioindicators. Agriculture, Ecosystems & Environment. 1999;**74**(1-3):137-115

[158] Pelosi Pelosi C, Joimel S, Makowski D. Searching for a more sensitive earthworm species to be used in pesticide homologation tests–a meta-analysis. Chemosphere. 2013;**90**(3):895-900

[159] Yasmin S, D'Souza D. Effects of pesticides on the growth and reproduction of earthworm: A review. Hindawi Publishing Corporation. Applied and Environmental Soil Science. 2010;**2010**(3):9. DOI: 10.1155/2010/678360

[160] Muñoz-Leoz B, Ruiz-Romera E, Antigüedad I, Garbisu C. Tebuconazole application decreases soil microbial biomass and activity. Soil Biology and Biochemistry. 2011;**43**(10):2176-2183

[161] Tripathi S, Srivastava P, Devi RS, Bhadouria R. Influence of synthetic fertilizers and pesticides on soil health and soil microbiology. In: MNV P, editor. Agrochemicals Detection, Treatment and Remediation: Pesticides and Chemical Fertilizers. Oxford, UK: Butterworth-Heinemann; 2020. pp. 25-54

[162] Tien CJ, Chen CS. Assessing the toxicity of organophosphorus pesticides to indigenous algae with implication for their ecotoxicological impact to aquatic ecosystems. Journal of Environmental Science and Health, Part B. 2012;**47**(9):901-912

[163] Rani M, Shanker U, Jassal V. Recent strategies for removal and degradation of persistent & toxic organochlorine pesticides using nanoparticles: A review. Journal of Environmental Management. 2017;**190**:208-222

[164] Kafilzadeh F. Assessment of organochlorine pesticide residues in water, sediments and fish from Lake Tashk, Iran. Achievements in the Life Sciences. 2015;**9**(2):107-111

[165] Simon-Delso N, Amaral-Rogers V, Belzunces LP, Bonmatin JM, Chagnon M, Downs C, et al. Systemic insecticides (neonicotinoids and fipronil): Trends, uses, mode of action and metabolites. Environmental Science and Pollution Research. 2015;**22**(1):5-34

[166] Germany PAN. Pesticide and Health Hazards: Facts and Figures. Hamburg, Germany: PAN Germany—PestizidAktions-Netzwerk eV; 2012

[167] Nicolopoulou-Stamati P, Maipas S, Kotampasi C, Stamatis P, Hens L. Chemical pesticides and human health: The urgent need for a new concept in agriculture. Frontiers in Public Health. 2016;**4**:148

[168] Kalafati L, Barouni R, Karakousi T, Abdollahi M, Tsatsakis A. Association of pesticide exposure with human congenital abnormalities. Toxicology and Applied Pharmacology. 2018;**346**:58-75

[169] Kalliora C, Mamoulakis C, Vasilopoulos E, Stamatiades GA, Scholtz MT, Voldner E, et al. A pesticide emission model (PEM). Part I: Model development. Atmospheric Environment. 2002;**36**:5005-5013

[170] Mostafalou S, Abdollahi M. Pesticides and human chronic diseases: Evidence, mechanisms, and perspectives. Toxicology and Applied Pharmacology. 2013;**268**(2):157-177

[171] Agrawal A, Sharma B. Pesticides induced oxidative stress in mammalian systems. International Journal of Biological and Medical Research (IJBMR). 2010;**1**(3):90-104

[172] Shaikh S, Fatima J, Shakil S, Rizvi SMD, Kamal MA. Antibiotic resistance and extended spectrum beta-lactamases: Types, epidemiology and treatment. Saudi Journal of Biological Sciences. 2015;**22**(1):90-101

[173] Ali S, Hameed S, Shahid M, Iqbal M, Lazarovits G, Imran A. Functional characterization of potential PGPR exhibiting broad-spectrum antifungal activity. Microbiological Research. 2020;**232**:126389

[174] Berg G. Plant–microbe interactions promoting plant growth and health: Perspectives for controlled use of microorganisms in agriculture. Applied Microbiology and Biotechnology. 2009;**84**(1):11-18

[175] Podile AR, Kishore GK. Plant growth-promoting rhizo-bacteria. In: Gnanamanickam SS, editor. Plant-Associated Bacteria. Netherlands: Springer; 2006. pp. 195-230

[176] Ngumbi E, Kloepper J. Bacterial-mediated drought tolerance: Current and future prospects. Applied Soil Ecology. 2016;**105**:109-125

[177] Ulloa-OgazAL,Munoz-CastellanosLN, Nevarez-Moorillon GV. Biocontrol of phytopathogens: Antibiotic production as mechanism of control, the battle against microbial pathogens. In: Mendez Vilas A (ed) Basic Science, Technological Advance and Educational Programs. vol. 1. Badajoz: Formatex Research Center; 2015. pp. 305-309

[178] Jayaprakashvel M. Mathivanan N. Management of plant diseases by microbial

metabolites. In: Bacteria in Agrobiology: Plant Nutrient Management. Berlin, Heidelberg: Springer; 2011. pp. 237-265

[179] Liu Y, Chen Z, Ng TB, Zhang J, Zhou M, Song F, et al. Bacisubin, an antifungal protein with ribonuclease and hemagglutinating activities from Bacillus subtilis strain B-916. Peptides. 2007;**28**(3):553-559

[180] Conrath U, Beckers GJ, Langenbach CJ, Jaskiewicz MR. Priming for enhanced defense. Annual Review of Phytopathology. 2015;**53**:97-119

[181] Van Loon LC. Induced resistance in plants and the role of pathogenesis-related proteins. European Journal of Plant Pathology. 1997;**103**:753-765

[182] Ahn IP, Lee SW, Suh SC. Rhizobacteria-induced priming in Arabidopsis is dependent on ethylene, jasmonic acid, and NPR1. Molecular Plant-Microbe Interactions. 2007;**20**(7):759-768

[183] Bhattacharyya PN, Jha DK. Plant growth promoting rhizobacteria (PGPR): Emergence in agriculture. World Journal of Microbiology and Biotechnology. 2012;**28**:1327-1350

[184] Prathap M, Kumari BR. A critical review on plant growth promoting rhizobacteria. Journal of Plant Pathology and Microbiology. 2015;**6**(4):21-319. DOI: 10.4172/2157-7471.1000266

[185] Pieterse CM, Zamioudis C, Berendsen RL, Weller DM, Van Wees SC, Bakker PA. Induced systemic resistance by beneficial microbes. Annual Review of Phytopathology. 2014;**52**:347-375

[186] Naznin HA, Kimura M, Miyazawa M, Hyakumachi M. Analysis of volatile organic compounds emitted by plant growth promoting fungus phoma sp. GS8-3 for growth promotion effects on tobacco. Microbes and Environment. 2012;**28**:42-49

[187] Solano BR, Maicas JB, Mañero FG. Physiological and molecular mechanisms of plant growth promoting rhizobacteria (PGPR). In: Plant-bacteria Interactions-Strategies and Techniques to Promote Plant Growth. Weinheim: Wiley VCH; 2008. pp. 41-54

[188] Pieterse CM, Van der Does D, Zamioudis C, Leon-Reyes A, Van Wees SC. Hormonal modulation of plant immunity. Annual Review of Cell and Developmental Biology. 2012;**28**:489-521

[189] Doornbos RF, Van Loon LC, Peter AHM, Bakker A. Impact of root exudates and plant defense signaling on bacterial communities in the rhizosphere. Renewable & Sustainable Development. 2012;**32**:227-243

[190] Meena VS, Maurya BR, Meena SK, Meena RK, Kumar A, Verma JP, et al. Bacillus species enhance nutrient availability in agricultural soils? In: Bacilli and Agrobiotechnology, 367-395. Cham: Springer; 2016

[191] Joshi M, Shrivastava R, Sharma AK, Prakash A. Screening of resistant varieties and antagonistic Fusariumoxysporum for biocontrol of fusarium wilt of Chilli. Plant Pathology and Microbiology. 2012;**3**:134

[192] Goswami D, Thakker JN, Dhandhukia PC. Portraying mechanics of plant growth promoting rhizobacteria (PGPR): A review. Cogent Food & Agriculture. 2016;**2**(1):1127500

[193] Nadeem SM, Naveed M, Zahir ZA, Asghar HN. Plant-microbe interactions for sustainable agriculture: Fundamentals and recent advances. In: Arora NK, editor. Plant Microbe

Symbiosis: Fundamentals and Advances. India: Springer; 2013. pp. 51-103

[194] Upadyay SK, Maurya SK, Singh DP. Salinity tolerance in free living plant growth promoting rhizobacteria. Indian Journal of Scientific Research. 2012;**3**:73-78

[195] Rabeendran N, Moot DJ, Jones EE, Stewart A. Inconsistent growth promotion of cabbage and lettuce from Trichoderma isolates. New Zealand Journal of Plant Protection. 2000;**53**:143-146

[196] Ryu CM, Farag MA, Hu CH, Reddy MS, Kloepper JW, Pare PW. Bacterial volatiles induce systemic resistance in Arabidopsis. Plant Physiology. 2004;**134**:1017-1026

[197] Tahir HA, Gu Q, Wu H, Raza W, Hanif A, Wu L, et al. Plant growth promotion by volatile organic compounds produced by Bacillus subtilis SYST2. Frontiers in Microbiology. 2017;**8**:171

[198] Naseem H, Ahsan M, Shahid MA, Khan N. Exopolysaccharides producing rhizobacteria and their role in plant growth and drought tolerance. Journal of Basic Microbiology. 2018;**58**:1009-1022. DOI: 10.1002/jobm.201800309

[199] Schalchli H, Tortella GR, Rubilar O, Parra L, Hormazabal E, Quiroz A. Fungal volatiles: An environmentally friendly tool to control pathogenic microorganisms in plants. Critical Reviews in Biotechnology. 2016;**36**(1): 144-152

[200] Tahir HAS, Gu Q, Wu H, Raza W, Safdar A, Huang Z, et al. Effect of volatile compounds produced by Ralstoniasolanacearum on plant growth promoting and systemic resistance inducing potential of bacillus volatiles. BMC Plant Biology. 2017;**17**(1):1-16

[201] Rudrappa T, Biedrzycki ML, Kunjeti SG, Donofrio NM, Czymmek KJ, Paul WP, et al. The rhizobacterial elicitor acetoin induces systemic resistance in Arabidopsis thaliana. Communicative & Integrative Biology. 2010;**3**(2):130-138

[202] Kanchiswamy CN, Malnoy M, Maffei ME. Chemical diversity of microbial volatiles and their potential for plant growth and productivity. Frontiers in Plant Science. 2015;**6**:151

[203] D'Alessandro D'Alessandro M, Erb M, Ton J, Brandenburg A, Karlen D, Zopfi J, et al. Volatiles produced by soil-borne endophytic bacteria increase plant pathogen resistance and affect tritrophic interactions. Plant, Cell & Environment. 2014;**37**:813-826. DOI: 10.1111/pce.12220

[204] Blumer C, Haas D. Mechanism, regulation, and ecological role of bacterial cyanide biosynthesis. Archives of Microbiology. 2000;**173**:170-177

[205] Lanteigne C, Gadkar VJ, Wallon T, Novinscak A, Filion M. Production of DAPG and HCN by Pseudomonas sp. LBUM300 contributes to the biological control of bacterial canker of tomato. Phytopathology. 2012;**102**(10):967

[206] Reetha S, Bhuvaneshwari G, Thamizhiniyan P, Mycin TR. Isolation of indole acetic acid produces bacteria of Pseudomonas florescens and Bacillus subtilis and enhances the growth of onion (Allimcepa. L). International Journal of Current Microbiology and Applied Sciences. 2014; **3**(2):568-574

[207] Dixit R, Agrawal L, Gupta S, Kumar M, Yadav S, Chauhan PS, Nautiyal CS. Southern blight disease of tomato control by 1-aminocyclopropane-1-carboxylate (ACC) deaminase

producing Paenibacilluslentimorbus B-30488. Plant Signaling & Behavior. 2016; **11**(2), e1113363

[208] Sundh I, Goettel MS. Regulating biocontrol agents: A historical perspective and a critical examination comparing microbial and macrobial agents. BioControl. 2012;**58**:575-593

Section 2

Impact on Medicinal and Aromatic Plants

Chapter 4

Using Biostimulants Containing Phytohormones to Recover Hail-Damaged Essential Oil Plants

Zenzile Peter Khetsha, Moosa Mahmood Sedibe, Rudolph Johannes Pretorius, Phoka Caiphus Rathebe and Karabelo Moloantoa

Abstract

Hail can cause significant damage to aromatic and medicinal plants; however, this has never been investigated scientifically on most of aromatic and medicinal plants. Globally, essential oil crop producers primarily make use of agricultural crop insurance and costly mitigation strategies to recover lost production costs and alleviate hail-damaged plants. However, most aromatic and medicinal plants are not covered under agricultural crop insurance, and many commercial farmers are not able to regularly employ expensive alternative strategies. Therefore, hail damage may present a challenge to essential oil growers. The use of natural and synthetic phytohormones in a form of biostimulants, as an alternative biological mitigation strategy against hail damage in essential oil crops, has not received much attention, and there is no information on this topic. Exogenous applications of natural and synthetic biostimulants have consistently demonstrated growth enhancement, nutrient acquisition, yield and quality optimization, as well as physiological efficiency in plants. Biostimulants in a form of phytohormones are involved in diverse plant physiological processes, including the regulation of gene expression for adaptive responses to biotic and abiotic stresses. Using biostimulants, this chapter will detail the potential recovery response of aromatic and medicinal plants to hail damage, and the response of plants treated with biostimulants.

Keywords: biostimulants, phytohormones, post hail damage, recovery, secondary metabolites

1. Introduction

Prolonged droughts, increased floods and frequent extreme weather events are evidence of climate change, as a consequence of global warming [1]. Agriculture is adapting to the variability in global climatic conditions, with farmers continually developing strategies to respond to changing weather patterns [2]. Aromatic and medicinal plants are among those plants which are negatively affected by climate

IntechOpen

change [3]. In addition, there is a concern over climate change affecting the secondary metabolites of many medicinal and aromatic plants [4]. Hailstorms are one of the most common global natural disasters [5], and temperate zones seem especially prone to hailstorm events [6]. To date, there is no recorded data for the impact of hail damage on essential oil plants; however, it was reported that hail causes substantial damage to aromatic plants [7, 8].

Changes in the secondary metabolites of essential oil plants vary depending on the type of damage incurred by the plant [9]. Hail wounding on these plants affect specific aromatic and therapeutic attributes that make these plants economically important [10]. It has also been shown that hail damage, mechanical damage, and insect herbivory cause essential oil compound changes [10]. For example, hail damage simulation through leaf puncturing resulted in reduced menthone levels and increased pulegone concentrations in muña (*Minthostachys mollis* [Kunth] Griseb.). This subsequently altered the composition of the volatiles released from the damaged leaves [9].

A common mitigation strategy for the loss in yield caused by hail damage is crop insurance [11]. Globally, crop insurance against hail damage can be purchased for most commercial plants, but not for essential oil crops [12]. Other alternative strategies include the construction of hail nets as a preventative measure against hail damage. However, this strategy is often unfeasible as the high construction and maintenance costs require more herbage material to produce sufficient essential oil to recover these costs. This, in turn, requires more land for production. Some farmers increase the application of nitrogen after hail to facilitate the formation of new leaves and buds [13]. It has been reported that increased nitrogen fertilization increases rose geranium herbage material [14]; however, the essential oil quality is reduced (based on the ISO standard) [15]. In temperate regions where hail frequencies are high, agro-meteorologists implement strategies, such as hail forecasting and cloud seeding, to reduce the extent of hail damage [11].

Hail is a natural hazard that can cause significant loss to crop yields [16]. Hailstones larger than 8.4 mm in diameter can result in defoliation, which, in turn, initiates cell division and the synthesis of cellular components [6]. Such wounding can also lead to stress that affects plant growth and metabolic activities [10]. To produce a stress-response and recover metabolic functions, plants rely on the crosstalk between phytohormones [9]. Biostimulants are also used in the agricultural industry to mitigate against these types of biotic and abiotic stresses [17].

Plants perform unique functions in plant development and stress repair, as well as improving the primary and secondary metabolite content of plants, which directly affects the essential oil biosynthesis and quality [18]. Exogenous applications of natural and synthetic biostimulants have consistently demonstrated growth enhancement, yield and quality optimization, as well as physiological efficiency in plants [19, 20]. Phytohormones are involved in diverse plant physiological processes, including the regulation of gene expression for adaptive responses to biotic and abiotic stresses [21]. Primary biostimulants include auxins, abscisic acid (ABA), cytokinins (CKs), gibberellic acids (GAs) and ethylene [22]. Secondary biostimulants include jasmonates and its analogues (methyl jasmonate, MeJA), brassinosteroids (BRs), salicylic acid, polyamines, sterols, and dehydrodiconiferyl alcohol glucosides [22]. There is currently no specific hail-mitigation strategy for most medicinal and aromatic crops, and the effects of potential hailstorm damage on essential oil and aromatic plants, as well as mitigation and control mechanisms, require further investigation by researchers.

2. Extent of defoliation, and hail damage on plants

Extreme climatic conditions can lead to significant losses in the agricultural sector [2]. These losses have drastically increased, by at least 400% from 1960 to 2005 [5, 23, 24]. Hailstorms are one of the most common global natural disasters [5], and temperate zones seem especially prone to hailstorm events [6]. The following review focusses on hailstorm as one of the natural disasters responsible for crop loss and damage.

Hail is defined as precipitation in the form of small pellets, or pieces of ice, which vary in size from 0.5 mm to 100 mm [25]. Hail can cause extensive damage to growing crops and other vegetation [26]. Hail formation takes place in elongated convective cumulonimbus clouds, which are often associated with thunderstorms [6]. Severe hail-related losses result from hailstones above H2 size [27]. Hailstones above H2 cause defoliation, tearing, bruising, breaking, and the loss of stems and flowers [26]. Such wounding can also lead to plant stress, which affects growth and metabolic activities [28]. This can further predispose plants to disease outbreaks since wounds provide an entry point for plant pathogens [28].

The extent of hail damage depends on several factors, such as the number of hailstones per unit area, wind velocity [6], and hailstone size [29]. Changnon [29] and Fernandes et al. [24], reported that hail damage differs extensively among plant species, and is influenced by several factors, such as plant height, and leaf and stem morphology. Certain plant species, e.g. soybean (*Glycine max* [L.] Merr.], tea (*Camellia sinensis* [L.] Kuntze), and tobacco (*Nicotiana tabacum* L.), are susceptible to damage by hailstones of any size (>H0), due to their leaf and stem morphology [29, 30]. In contrast, crops such as maize and wheat are mostly damaged by larger, windblown hailstones (>H3) because of their height, and differences in leaf area, the stem sizes and morphology [29, 30].

Plants also respond differently to wounding stress following hail damage. Physical wounding of plant tissue resulting from hail damage and defoliation initiates a cascade of biochemical or physiological processes, which results in the repair of damaged tissue and resistance to opportunistic pests and pathogens [31]. Such alterations occur both in the tissue immediately surrounding the wound, and in distal tissue not directly in contact with the damaged tissue [31]. At the wound site, cell division and the synthesis of cellular components that are required to isolate the damaged tissue, reduce water loss, and restore tissue integrity, is initiated [31]. In leaves, wounding also induces the synthesis and accumulation of anti-microbial compounds, and in the specific case of damage caused by herbivory, volatile metabolites are released to deter pests and attract their predators and parasitoids [31].

Hail damage hinders plant growth, and affects the yield and quality by changing the cellular metabolic processes [4]. Defoliation and wounding stress in plants results in a knock-back effect that reduces the assimilation of carbon, ultimately affecting the rate of photosynthesis [4]. This stress stimulates the production of bio-inhibitors, reactive oxygen species levels, transient Ca^{2+} influxes into the cytoplasm, and protein phosphorylation. It also causes irreversible injury to cells and tissue that eventually slows growth [32]. This has been reported in potato plants (*Solanum tuberosum* L.), where ribonucleic acid (RNA) homology changed as a result of wounding [33]. Christopher et al. [34] identified a suite of wound-regulated genes, indicating the diversity and multiplicity of the induced defense response in systemically wounded leaves of hybrid poplars (*Populus trichocarpa* × *P. deltoides*). In lettuce (*Lactuca sativa* L.), plant wounding induced the synthesis of phenylalanine ammonia-lyase compounds [35].

The local and systemic plant responses activate and regulate defense mechanisms for localized tissue damage, such as those resulting from hail damage [36]. Plants can also positively adapt with altered growth habits to contradict the damaging effects of hail [4]. Thus, the responses are both reversible and irreversible modifications, such as cell division, alterations of membrane channels, and a change in the structure of the cell wall [4]. This has been demonstrated with muña (*Minthostachys mollis* [Kunth] Griseb.), where leaf puncturing resulted in reduced menthone levels, while the pulegone concentration increased during the first 48 h of the experiment [9]. Furthermore, an increased pulegone level and diminishing menthone emissions also altered the composition of volatiles released from damaged leaves. Depending on the phytohormonal crosstalk and proteins released, wound-inducible genes may either repair damaged plant tissue or produce inhibitory system [36].

3. Biostimulants as a mitigating strategy for hail damaged essential oil plants

Phytohormones are molecules that influence the growth and development of plants, even at low concentrations. There are phytohormones which are produced naturally by plants, as well as synthetic regulators, which are have been produced in biotechnology research as biostimulants [9]. Biostimulants are used in global crop production to improve field competitiveness, nutrient use efficiency, and stress resistance in plants [37]. Most biostimulants and their products are extracted from algae, arbuscular mycorrhizal fungi, chitin and chitosan derivatives, fulvic and humic acids, synthetic phytohormones, natural plant extracts and the magnetic field as a growth stimulant for plant species [37, 38]. Exogenous applications of natural and synthetic biostimulants enhance growth, increase oil yield and quality, as well as improving physiological efficiency in plants [37]. This section focuses on the use of biostimulants to recover defoliated, wounded, and hail-damaged plants.

Plant metabolic activities are regulated by phytohormones [39]. Phytohormones are produced naturally by plants and are small organic molecules which, at low endogenous concentrations (10^{-6} M to 10^{-9} M), induce metabolic activities within plant cells to modify growth and development [22]. However, synthetic chemicals, with the same properties and effects as natural phytohormones, can be produced [9]. A key research objective in plant biotechnology is to recognize the mechanism used by plants to respond to and overcome different environmental stressors [4]. Phytohormones are classified as either plant growth regulators (CK, GA, BRs, and auxins) or as bio-inhibitors (ABA, jasmonates, and salicylic acid) [20]. Phytohormones are involved in a number of diverse plant physiological processes, including the regulation of gene expression for adaptive responses to biotic and abiotic stress [21] and in wound healing [4, 17].

Primary phytohormones include auxins, ABA, CK, GA, and ethylene [40]. Other naturally occurring phytohormonal-like molecules include jasmonates, BRs, salicylic acid, polyamines, sterols, and dehydrodiconiferyl alcohol glucosides [40]. Phytohormones alter metabolic activities associated with cell division, cell enlargement, flowering, fruiting, and seed formation. Nemhauser et al. [19] found that exogenous applications of phytohormones regulate plant growth, and greatly influence plant stature and organ size. Bio-inhibitors are involved in the alleviation of biotic and abiotic stress that results from wounding, moisture stress, and temperature stress [20].

3.1 Effects of phytohormones, biostimulants and magnetic field on plant growth and development, refoliation and wounding

Phytohormones promote growth and development, the production of secondary metabolites, as well as bio-inhibition, due to the extensive crosstalk and signal integration which affects the plants physio-morphological chemistry [9]. The recovery mechanism of defoliated, wounded, and hail-damaged plants has provided the rationale for investigating the potential of phytohormone application in this chapter. The following section examines the effects of natural and synthetic ABA, jasmonates, BRs, CK and GA on plant growth and development, and changes in primary and secondary metabolite biosynthesis of essential oil plants.

3.1.1 Abscisic acid as a potential growth stimulant for refoliation and wounding recovery

Abscisic acid is a stress-signaling molecule, found to occur in all kingdoms, except Archea. The effect of foliar applications of ABA on plant growth is dependent on the plant species [41]. It is a crucial regulator of important plant processes, including resilience to abiotic stressors, such as wounding, moisture, light, drought, and temperature [42]. Abscisic acid is also involved in primary plant growth development. This includes buffering the day-night alterations of leaf growth rate, and regulating stomatal movement and transpiration rate [43]. It also improves leaf growth by increasing tissue and whole-plant hydraulic conductivity [44]. Dammann et al. [45] found that hail damage, and defoliation causes increased levels of ABA in plants, which in turn activates the biosynthesis of jasmonates.

Abscisic acid is a C15 sesquiterpenoid, formed by the joining of three isoprenoid units [4, 41]. Abscisic acid occurs naturally as (S)-(+)-ABA, which is often called a cis isomer; a combination of 1:1 *cis* and *trans* abscisic acids, classed as isoprenoids [44]. The ABA biosynthetic pathway starts from oxidative cleavage of epoxy-carotenoids and 9-cis violaxanthin, where xanthoxin is converted to abscisic acid [4, 41]. The early C5 precursor of ABA, isopentenyl pyrophosphate, is produced primarily in plastids, via 1-deoxy-D-xylulose-5-phosphate from pyruvate and glyceraldehydes-3-phosphate. This is then processed to farnesyl pyrophosphate, geranylgeranyl pyrophosphate, phytoene, carotene, and lycopene [4, 41]. Subsequently, xanthoxin migrates from the plastid to the cytosol, where it is converted to ABA by abscisic aldehyde, xanthosis acid, or abscisic alcohol; clearly, abscisic aldehyde is an intermediary in the conversion of xanthoxin to abscisic acid [4, 41].

Transportation of ABA primarily occurs through the vascular tissues of plants [46]. However, ABA responds to abiotic stress through the cells [41]; this requires translocation from ABA-producing cells, via intercellular transport, to allow rapid distribution to other plant tissue [41]. Abscisic acid is ubiquitous in plants; the endogenous levels in plant cells determine its homeostasis [4]. However, developmental and environmental factors such as light, wounding, salinity, and water stress affect ABA concentration levels [4].

Abscisic acid is sensitive to direct sunlight and high temperatures, and rapidly deteriorates under these conditions [47]. Kong and Zhao [48] found that foliar-applied ABA elevates the aroma content of aromatic rice (*Oryza sativa* L). Foliar application of ABA on yeba mate (*Ilex paraguariesnsis* A.St.-Hil.) increases plant growth, by reducing water stress through stomatal closure [49]. It has also been reported that the exogenous application of ABA improves wounding tolerance in tomatoes and potatoes, and induces jasmonate gene expression in the leaves and roots of potatoes [49]. The exogenous application of ABA should preferably be conducted during cool mornings, before sunrise and increasing temperatures.

As well as direct involvement in plant growth and development, ABA has a significant role in the regulation of environmental stress. [50] reported that there are two distinct pathways responsible for the developmental and environmental stress regulatory processes. Responses to plant wounding includes the local response at the site of the wound, and the systemic response, which occurs throughout the whole plant [36]. The foliar concentrations of ABA applied to wounded plants varies from 0.001 mM to 1.5 mM solution per plant, with the optimal application rate and concentration differing between different crop species [48, 49, 51, 52].

3.1.2 Jasmonates as a potential growth stimulant for refoliation and wounding recovery

Jasmonates and their methyl ester, MeJA, are natural-occurring growth regulators found in higher plants [53]. Jasmonates were first discovered and isolated from a culture of the fungus *Lasiodiplodia theobromae* (Pat.), while MeJA is a component of the essential oils of *Jasminum grandiforum* (L.) and *Rosmanus officinalis* (L.) [53]. Cyclopentanone-derived jasmonates are widespread in plants [54], while MeJA, and jasmonic acid and its hydroxylated derivatives are commonly used jasmonates for agricuclural purposes. In plants, cis-jasmone, jasmonyl-1-aminocyclopropane-1 carboxylic acid, and jasmonoyl isoleucine are also known to act as analogues of jasmonates [54].

Jasmonates are present throughout the plant body, with the highest concentration in growing tissues such as shoot tips, root tips, immature fruits, and young leaves [54]. Jasmonate biosynthesis in plants ranges from 0.01 μg/g to 3 μg/g in fresh mass [54]; however, it has been found to be as high as 95 μg/g in fresh mass of sagebrush (*Artemisia tridentate* [Nutt.]) [53]. Jasmonates are derived from the fatty acid metabolism pathway and are harvested directly from jasmine (*Jasminum grandiflorum* L.) [53]. Huang et al. [55] reported that biosynthesis of jasmonates takes place in three different cell membranes, and is activated by 13-lipoxygenase to form hydroperoxyoctadecatrienoic acid.

Jasmonates, along with their derivatives, control various aspects of plant growth and development, such as stamen development, root development, flowering, and leaf senescence [55]. Jasmonates also induce a variety of physiological processes, such as seed germination, pollen development, ethylene synthesis, tuber formation, fruit ripening, and tendril coiling [56]. However, when applied exogenously, jasmonates can modulate stress by either enhancing or suppressing plant development [57].

Jasmonates also activate a signal transduction pathway in response to different kinds of stress [57]. Plant responses to abiotic stresses, such as wounding, are coordinated both locally and systemically by jasmonate signaling molecules [56]. In addition, there is a causal link between wounding and jasmonates; wounding causes the release of linoleic acid, a jasmonate precursor, from the membrane lipids, in turn forming jasmonate [56]. Thus, the jasmonate signaling pathway involves signal transduction events that are regulated by wounding shock, especially in relation to leaf defoliation [56].

The molecular mechanism of crosstalk between growth and immune-signaling networks are regulated and mediated through biosynthetic pathways of phenylpropanoids, polyketides, terpenoids, and N-containing compounds, which are directly associated with jasmonates [58]. The effects of exogenously applied jasmonic acid on growth, changes in essential oil biosynthesis, and plants subjected to biotic and abiotic stresses has been tested [59, 60]. In a study conducted by [61], a high concentration of MeJA (1 mM) inhibited primary root growth of soybeans, while a low application concentration (0.01 μM) slightly stimulated root growth. Anderson [62] also reported that low levels (1 μM–10 μM) of MeJA alters protein and mRNA populations, without inducing senescence in cell culture, while a high concentration of jasmonic acid or MeJA (50 μM) induces senescence in cell culture, and slows the primary root growth of soybeans [61].

Methyl jasmonate applied at 0.5 mM increases the content of eugenol and linalool in basil plants compared to the control [63]. In another study, MeJA applied to bigleaf marsh-elder (*Iva frutescens* L.) resulted in an increase in volatile compounds (α-pinene, sabinene, and limonene emissions) [64]. The increase in volatile emissions following MeJA treatment was ascribed to terpene synthase activation and de novo synthesis. In these studies, foliar application of MeJA varied between 0.01 mM and 0.5 mM solution, per plant, with the application rate and concentration differing between different crop species, under normal growing conditions [65–68].

3.1.3 Combined cytokinin and gibberellin, and the brassinosteroids as potential growth stimulants for recovery of defoliated and hail-damaged essential oil plants

The effects of phytohomones are based on the synergism to improve the growth and development, as well as the recovery, resistance and survival of stressed plants [60, 69]. Phytohormones rarely function independently; they depend on a cross-talk network between their synergic and antagonistic metabolic processes [40]. Gibberellins primarily controls cell growth and division by stimulating the elongation of internodes [70]. In a study by [71], longer internodes and delayed flowering were observed in tomatoes treated with GA (5 mg/L), while plants treated with only CK (5 mg/L) formed no axillary buds. However, when GA and CK were combined, there was an increase in fresh matter [71]. This demonstrates that there is a possible interaction between the major biostimulants, with antagonistic relations, which improves plant growth and development. The crosstalk between GA and CK involves components from the GA biosynthesis pathways, which plays a central role in the regulation of plant growth and development [72]. Gibberellins and CKs are commonly used in agriculture, viticulture, gardens, and horticulture [73].

Cytokinin was first discovered in the early 1940s when coconut milk was added to aid cell division in tobacco plants [74]. All CKs are adenine derivatives and mostly occur as either free compounds, glucosides, or ribosides in the plant root system, particularly the root apex [44, 75]. According to [44], CK biosynthesis occurs through biochemical modification of dimethylallyl diphosphate, which is initiated through the transference of the isopentenyl moiety from dimethylallyl diphosphate to the N6 position of adenosine triphosphate, catalyzed by isopentenyl transferases. These form the isopentenyl transferases and the isoprene side chain, which is subsequently trans-hydroxylated by cytochrome P450 (CYP450) to yield zeatin ribosides [44]. The metabolic storage and transport of CK is not yet fully understood, however, it is hypothesized that transport takes place via the vascular tissue (particularly the xylem), from the roots to the shoots of the plant [76].

Commonly-used CKs for agricultural purposes include zeatin, kinetin, 6-Benzylaminopurine (BA and BAP), 2-isopentenyl adenine, zeatin riboside, and dihydro-zeatin [40, 77]. The main functional properties of CKs for agricultural use are the stimulation of cell division, release of lateral bud dormancy, the induction of adventitious bud formation, retarded leaf senescence, and the promotion of chlorophyll synthesis [78]. Exogenous application of CKs is currently used to optimize the internal concentrations of CKs for growth and development, organ regeneration after wounding damage, and changing the chemical compositions of essential oils [78].

Gibberellins were first isolated from the fungus, *Fusarium moniliformae* Sheldon (*Gibberella fujiskuroi* [Swada] Wollenweber), and named gibberellic acid (GA_3) [79]. To date, ca. 126 naturally-occurring GAs have been discovered [73], with each plant species containing at least six to ten GAs [80]. A wise decision was made early in GA research to number the various GAs, rather than naming them separately, as was done with chemical-related sterols. Therefore, GAs are known as GA_1, GA_2, and GA_3, etc. up to more than GA_{126} [81].

Gibberellins are a large group of essential diterpenoid acids [73]. They are biosynthesized in shoot apices, young leaves, and flowers of plants, via the terpenoid pathway [81]. Biosynthesis of GA requires three enzymes viz., terpene synthase, CYP450s, and 2-oxoglutarate [81]. Gibberellins are transported through plants by means of the vascular tissues, xylem, and phloem [81].

The most common molecular mechanisms of GA signaling in plants is through the GA receptor, Gibberellin Insensitive Dwarf 1 (*GID1*) [44, 82]. Upon binding, the receptors undergo conformation changes that favor the binding of *DELLA* proteins, a group of nuclear transcriptional regulators that repress GA signaling and plant growth [44, 82]. Homeostasis and regulation of the GA biosynthetic pathway depends on the developmental and environmental signals, specifically the genes GA_{20}-oxidases, GA_3-oxidase, and GA_2-oxidases [44].

Gibberellins are endogenous hormones functioning as biostimulants that influence a wide range of developmental processes in higher plants [72]. This includes plant growth and development through promoting leaf development, stem elongation, induction of seed germination, promotion of hypocotyls and stem elongation, regulation of pollen development, and flower initiation [81]. Some GA-deficient mutants can cause dwarfism [82]. Different types of GAs are used to achieve specific agronomic objectives, for example, anti-flowering GA_7 and GA_3 are commonly used for promoting germination, seed development, leaf development, and stem elongation [81].

Several studies have revealed a reciprocal developmental dependence between the two hormones, where the ratio between GA and CK affects the developmental processes of the plant [72]. Cato et al. [71] observed positive synergic crosstalk between GA and CK in tomatoes, and [75] reported that a combination of BA and GA induced longer tomato shoots, under different abiotic stress conditions. High CK and low GA signals are required for normal shoot apical meristem functioning [82]. In contrast to these findings, joint applications of GAs and CKs have been shown to exert antagonistic effects on numerous developmental processes, including shoot and root elongation, cell differentiation, shoot regeneration in culture, and meristem activity [83, 84]. Moreover, GA tends to inhibit CK-induced cell differentiation in plants [72]. This inhibition is attributed to the loss of the *SPINDLY* protein function, which results from CK resistance.

Cytokinin activity is highest during early shoot initiation (controlling meristem activity) [85]; in contrast, GAs act at a later stage to regulate plant cell division and shoot elongation [84]. The GA biosynthetic pathway from trans-geranylgeranyl diphosphate to GA_{12}-aldehyde, leads to the identification of positive and negative signaling components [72]. In Arabidopsis (*Arabidopsis thaliana* L.), GA and CK signaling regulated the expression of ARR1 through repression of GA, via degradation of the *DELLA* protein RGA [86]. This indicates that reducing the GA concentration, 5 days after foliar application, releases *ARR1* from repression, which in turn upregulates *SHY2*. This leads to an increase in cell differentiation, which balances with cell division to control plant growth. In addition, the regulation of *SHY2* by *ARR1* also represents a point of crosstalk with auxin, thus connecting three hormones in a single network [86]. The homeostatic balance of GA and CK in plants may vary between species, and the response of plants to foliar treatments may differ as a result of external factors, such as change in environmental conditions, and stage of development [87].

Changes in the ratios of GAs and CKs to both each other, and to other hormones, often results in distinct and divergent morphological features, such as dwarfism, contorted or twisted growth, weeping forms, or fastigiated and columnar forms [89]. In addition, the ratios may cause extra-large leaves or elongated stems, and extensive shoot proliferation, especially under less favorable environmental conditions. Weeping forms have been observed in spruce (*Picea abies* Mill.), pine (*Pinus densiflora* L.), and sweet viburnum (*Viburnum odoratissimum* Ker Gawl.) treated with combined GA and CK [88, 89]. Therefore, it is evident that under abiotic stress conditions, plants can be treated with combined GA and CK to induce foliage material development. However, it is crucial to plan the application scheduling and ratios for each of these hormones, as shown in studies by [83, 84].

Brassinosteroids were named after the genus *Brassica* during the late 1970s [90], and they were initially extracted from maize pollen [90, 91]: to date, there are more than 70 free BRs and conjugates, described from various plant species [91]. Trace amounts of BR's in a complex plant matrix are classed as polyhydroxylated steroid plant hormones and are widespread among plants [91]. In addition, BRs are structurally classed as C27, C28, or C29 based on the different alkyl-substitution patterns of the side chains [92]. All the BRs isolated from plants are produced through campesterol biosynthesis and belong to the C28-BRs, with a 24α-methyl group [80].

The most effective BRs, which have been extracted from plants for agricultural use are brassinolide, castasterone, testosterone, and 6-deoxy castasterone [80]. Of all the BRs, brassinolide is biologically the most active [92]; it is ubiquitous in plants and is produced in almost all plant parts, where it controls growth and developmental processes [93]. Plants synthesize excess brassinolide to meet the continuous need for growth and development, while inactive brassinolide is converted into active forms to maintain BRs homeostasis [92].

Brassinosteroids are found in various plant species, including monoplast freshwater algae and brown algae, suggesting that they are ancient ubiquitous plant hormones [92]. Brassinosteroids are also found in pollen, immature seeds, roots, and flowers [91, 92]. They range from 1 to 100 ng/g fresh weight in flowers, while shoots and leaves have lower amounts of 0.01–0.1 ng/g fresh mass [91, 92]. Brassinosteroids are not mobile within plants, they function by paracrine or autocrine signaling; however, long-distance transport of exogenously-applied BRs does occur in plants, particularly from the roots to shoots, but foliar-applied BRs (24-epibrassinolde) are fixed in the leaves [92]. In addition, [94] reported high mobility of BRs in a plant system.

Brassinosteroids are involved in a wide variety of plant physiological activities. They regulate plant growth, at nanomolar to micromolar concentrations, for multiple developmental processes, including cell division, cell elongation, vascular differentiation, reproductive development, and modulation of gene expression [54]. High metabolic activity, associated with growth, has been observed in rape (*Brassica napus* L.) treated with BRs [90]. BRs are also involved in microbial infection recovery, hypocotyl growth, increased leaf lamina growth, increased shoot apex fresh weight, pollen tube growth, and stress tolerance [80].

The application of BRs enhances plant biomass, secondary metabolites, antioxidant defense activities, and the accumulation of osmoprotectants under biotic and abiotic stress [93]. This has been demonstrated in soybean, where the application of 1 μmol/L BRs led to hypocotyl and epicotyl elongation. However, epicotyl elongation was affected by photoperiod, with no increase in length under dark conditions [95]. Therefore, BRs applied at higher concentrations (≥1 μmol/L) in dark-grown plants suppress shoot and root development [95]. Mung beans treated with the BR, 28-homobrasinolide, at 10^{-8} M, had increased leaf area, and plant height, as well as fresh and dry mass of shoots and roots over a 21-day growth period. This treatment also increased proline content [96]. The foliar fresh matter of corn mint (*Mentha arvensis* L.f. *Piperascens Malinv*. Ex Holmes) and its menthol content increased when treated with lactonic spirostane-SABS (0.5 ng/L) and ketonic-SABS (0.5 ng/L) [97]. BRs induce chlorophyll synthesis, through the activation of enzyme proteins [98]. [99] demonstrated this, with a foliar application of 0.5 ppm BRs increasing the chlorophyll content of soybean.

In light of these studies, it is evident that BRs can have a significant impact on plant growth and development, and therefore on recovering the yield and essential oil quality parameters of defoliated, wounded, or hail-damaged plants. The role of BRs in the alleviation of various abiotic and biotic stressors, such as temperature, salinity, moisture, and heavy metal exposure has been reported [100].

3.1.4 Natural biostimulants-containing phytohormones as a recovery mechanism for wounded, defoliated and hail-damaged plants

Exogenous applications of synthetic biostimulants have been shown to consistently enhance growth, yield optimization and oil quality [60], as well as physiological efficiency in plants [19]. However, the production and availability of some synthetic phytohormones are expensive and not readily available. In addition, the practical use of these phytohormones is dependent on various environmental factors, such as temperature and light [101]. The use of synthetic biostimulants is a potential ecological hazard as they could pose a threat to the health of non-target organisms, especially when improperly used [90]. As such, the use of less harmful and cheaper bioactive stimulants are preferred over conventional synthetic phytohormones [80].

There are numerous commercially available bio-fertilizers, plant conditioners, allelopathic preparations, biogenic stimulators, elicitors, plant strengtheners, and biostimulants (**Table 1**). Most of these products are considered as biostimulants, containing biostimulants. However, [102] found that some biostimulants contain traces of natural phytohormones, but their biological action should not be attributed to them, unless registered as biostimulants.

Biostimulants products are developed based on the synergism between natural phytohormones [86, 87]. However, there are only a few published scientific reports on these products, since most industrial companies withhold information for market confidentiality purposes (examples in **Table 1**). Below are a few of the published reports to highlight the effects of registered biostimulants (containing traces of phytohormones) on plant growth, recovery, and resistance against stressors.

Application of biostimulants (undisclosed brand-name) containing GA (50 mg/L) and CK (90 mg/L) increased the number of leaves, flower heads and the total flavonoid content of marigold plants inflorescences [106]. Peppers (*Capsicum annuum* L.) treated with Megafol® (0.2% 40 mL; GA & CK content) had increased Ca uptake in

Product	Composition	Citation
Lucky Plant®	Gibberellic acid, BRs and traces of CKs	Agraforum, Germany
ComCat®	Brassinosteroids (2,4-epibrassinolide)	Agraforum, Germany
SilCat™	Brassinosteroids (2,4-epibrassinolide)	Agraforum, Germany
AnnGro® EW	Ethyl esters of fatty acids	Agraforum, Germany
Stimplex®	Cytokinin (Kinetin 0.01%)	[102]
Fungicidal	Salicylic acid	Pan African Farms
PanAf®	Salicylic acid	Pan African Farms
Megafol®	Amino acids, betaines, proteins, vitamins, auxin, GA, and cytokinins	[69]
Biozyme®	Algae extract, GAs, auxin and zeatin, and chelated micronutrients	[69]
Slavol	N-fixing, and phosphate-mineralizing bacteria, and auxins	[69]
Algreen, Leili®	Seaweed extract, plant growth regulator, vitamins, free amino acids, and alginic acid	[69]
Agrispon®	Natural plant extract with traces of phytohormones	[103]
Kelpak®	Seaweed extract and traces of plant growth regulators	[103]
Stifun	Complex biologically active substances of natural origin	[104]
Auxym®	Auxins and CK, amino acids, peptides, vitamins, and essential micronutrients	[105]

Table 1.
Examples of commercial biostimulants containing phytohormones as declared on the labels.

the foliage material and fruits. Megafol® also increased the fruiting yield and the level of antioxidants in the same species [69]. Hüster [90] studied the stimulatory effects of ComCat® on wheat, maize, cabbage (*Brassica oleracea* L.), carrots (*Daucus carota* [Hoffm.] Schübl.), onions (*Allium cepa* L.), lettuce, beetroot (*Beta vulgaris* L.), and peas (*Pisum sativum* L.). They demonstrated that an optimum foliar application of 5 mg/L, significantly increased the foliar biomass of these crops. Marjoram (*Majorana hortensis* Moench.) treated with BRs (25 mg/L) was rich in *cis*-sabinene hydrate content [107]. Foliar application of the tropical plant extract, Auxym® (2 ml/L), improved the yield of jute (*Corchorus olitorius* sp.) [105]. Auxym® also increased the chlorophyll content, enhanced the adaptation of jute plants to fluctuating light levels, and positively increased the starch, soluble proteins, and amino acids content when the plants received the full strength of nutrient solution [105]. From these studies, it is evident that phytohormones and biostimulants containing natural phytohormones can increase crop yield, enhance plant phytochemistry and secondary metabolites biosynthesis, and improve the recovery response mechanism of plants following biotic and abiotic stresses.

3.1.5 Magnetic field as a potential growth stimulant

The emission of magnetic fields (MFs) in the ecosystem due to ever changing and advancing technology has brought significant changes in the human and ecological environment [108]. In the modern days, the use of MFs as a stimulant for plant performance and capability has been far an interesting alternative method to chemical stimulants [109]. Magnetic fields have positively influenced the morphogenesis, showing great modification of seed germination, seedling growth and development in various plants such as cereals, grasses, medicinal plants, horticultural crops and herbs [38]. It is worth noting that MFs constitute non-toxic stimulus resulting in increased food and environmental safety. Application of MFs on crops has been seen to reduce the attack of pathogenic diseases [110]. Many studies have tried to understand the actual mechanism involved on how seeds germinate when exposed to MFs. Vashisth and Joshi [111] exposed seeds of maize to static-MFs for 4 h on strengths ranging from 50, 100, 150, 200 and 250 milliTesla (mT). The results suggested that MFs application enhance the seed percentage-germination, seedling length, dry weight and speed of germination when compared to the referent group. Furthermore, exposure to MFs reduced cellular leakage, improved water absorption and functional root parameters. Kirdan et al. [112] performed an interesting experiment by treating *Pinus Pinea* L. seeds with a MF of 9.42 mT for a different period of time; 0 min (used as a reference), 15, 30, and 40 min. Seeds exposed for 30 and 45 min showed a higher germination energy. There was also an increased root collar diameter, shoot height and tap root length.

Plants are outstanding experimental models compared to animals when conducting MFs exposure and response growth relationship studies. According to Vian et al. [113], they efficiently intercept with electromagnetic fields (EMFs) because of immobility and constant orientation. The benefit of magnetic seed germination has been seen in various biochemical events such as enzymatic stimulation, bioenergetic excitements and protein synthesis [110, 114]. Electrons in various molecules absorbs MF energy and utilize it for accelerating seed metabolism that triggers biochemical and enzyme reactions in the early stages of seed germination [115]. Afzal et al. [110] applied magnetic field strengths of 50, 100 and 150 mT for 5, 10 and 15 min on sunflower seed, and observed an increased α-amylase activity, with reduced sugar in high strength magnetically treated seeds. This confirms that magnetic treatment stimulates the protein synthesis of existing enzymes by producing germination metabolites at required amounts. Vashisth et al. [116] studied the effects of 200 mT for 2 h on crop growth and the yield of sunflower crops raised from magnetically treated seeds sown under different moisture stress conditions. The experimental results showed that plants from magnetically treated seeds had a higher leaf index area, chlorophyll content, 1000-seed mass, shoot length and biomass compared to untreated seeds. Magnetic field exposures in plants act as a stimulant in improving crop growth and yield under different ecological stress conditions.

3.2 Biostimulants, defoliation and hail damage on the development of leaf trichomes

The production and accumulation of essential oils is restricted to specialized structures (e.g. glandular trichomes, secretory cavities, and idioblasts) since they are toxic to healthy plant cells [9, 60]. The production of these essential oils takes place in closely connected secretory structure formations. It has been shown that biotic and abiotic stress factors affect essential oil production [9, 60]. In addition, plants produce some essential oil compounds in response to physiological stress, pathogen attack, and other ecological factors [60]. Therefore, it has a direct effect on the stimulation of the essential oil biosynthesis, which directly benefits essential oil yield and quality.

The recovery response mechanism of plants to hail damage, defoliation, wounding or grafting is complex, starting from upregulation of plant-stress hormones at the wound site, and later plant growth regulators to recover the lost organs [117, 118]. Therefore, the recovery of leaves following hail damage stress or related climate change affect the essential oil biosynthesis through the specialized structures called glandular trichomes, located on both surfaces of the leaf, and on tender stems and buds [52]. Moreover, [119] claimed that the densification of trichomes occurs as early as during leaf differentiation and continues throughout leaf development of *P. scabrum*.

Trichomes play different roles in plant physiology and ecology, especially with regards to morphological, mechanical and phytochemical characteristics [120]. Trichome density may vary with changes in environmental conditions [120]. These variations may indicate trade-offs between the trichomes, subsequently increasing resistance against trichome production. In defoliated plants, the rate of leaf regeneration is slow, possibly due to the endogenous ratio between CK and GA, and regulated plant bio-inhibitors [45, 121]. In the study conducted by [122], increasing the level of combined CK and GA decreased the density of the non-glandular trichomes on rose geranium, particularly on the adaxial leaf surface. In addition, [119] reported that non-glandular trichomes develop before glandular trichomes in the leaf primordial of *Pelargonium*, and cease development as the leaf expands. Therefore, this process could be ascribed to CK stimulation of the cell division, further regulating the leaf primordia by negatively affecting GA signaling through *IPT7* and GA_{20} oxidases, at an early stage of leaf primordia [123]. Zhou et al. [124] demonstrated this on Arabidopsis, where the C2H2 transcriptional factors regulated trichome cell differentiation through CK and GA pathways, suggesting that excess endogenous levels of these biostimulants could be toxic for trichome development. In another study, Zhou et al. [125] further reported that the *ZFP5* transcription factor of GA regulates trichome developmental actions, mainly through cell differentiation. In addition, GA is biosynthesized in young leaves and can translocate freely by diffusing through the plant cell protoplast when applied foliarly.

A combination of endogenous development and external signals regulate the developmental distribution of trichomes on plant leaves [126]. Therefore, under extreme external stimuli, such as complete defoliation, endogenous phytohormones are only synchronized to regenerate lost material. This directly affects leaf expansion and trichome developmental rate. In the study conducted by [122], the development and densities of the asciiform trichome following simulated hail damage was due to high concentration of GA (300 mg/L), applied at a later stage, in contrast to CK (0.64 mg/L) which was used earlier. Zhou et al. [124, 125] reported similar findings in Arabidopsis, where GA and CK, at concentrations as low as 100 μM, increased the density of glandular trichomes. According to the author, this was ascribed to GA and CK molecules, which regulated the development of glandular trichome through the combined action of *ZFP5* and *ZFP6* transcription factors [125]. Thus, in combination, these transcription factors regulate trichome cell differentiation, an important metabolic mechanism associated with trichome development.

Xue et al. [126] also demonstrated that the development of trichomes, and the biosynthesis of essential oil could be influenced by exogenous applications of BRs and jasmonic acid. According to [127], the *dpy* mutant (BR-deficient) is the one which enhances pubescence, while the jasmonic acid jai1-1 mutant produces the opposite phenotypic effect [127, 128]. The Arabidopsis *bls1* mutant, which is impaired in the BR response, developed fewer trichomes on both the abaxial and adaxial leaf surfaces, indicating the possible involvement of BR in trichome development [129].

Defoliated plants deploy stored resources to rebuild photosynthetic material and regenerate new organs or tissues following defoliation and wounding [118, 130]. During the refoliation, it is possible that the endogenous CK content in defoliated plants is already too high; then GA accumulation occurred at the later stage to regulate growth. According to [89], the amalgamation of CK and GA following hail damage may cause alterations in morphological features, such as increased trichome density. Other than GA and CK, BRs and jasmonic acid directly affect trichome formation through the accumulation of *zgb* and PI-I transcripts, indicating the importance of BRs in leaf recovery following defoliation or wounding [127].

3.3 Biostimulants, defoliation and hail damage on the production of primary and secondary metabolites

Plant chemistry (e.g. the essential oil and phytohormone content) is altered following mechanical damage, as caused by hail and/or animal herbivory [9]. Endogenous phytohormones are the primary inducible defense response for this class of volatiles, signaling to the transduction pathway between wounding stress perception and induction [36, 131]. These physiological response mechanisms occur within matter of minutes to several hours, resulting in the activation of wound-related defense genes [36].

According to [132], increased biomass and essential oil yield was recorded in rose geranium plants when biostimulants such as IAA, IBA, cycocel, cytozyne, biomyze, thephon, mepiquat chloride, triacontanol, and mixtalol were applied. However, the use of most of these natural or synthetic biostimulants as a recovery method for hail-damaged plants has not been tested. In addition, changes in the essential oil yield and quality have been found in most essential oil and medicinal plants, as a result of the exogenous application and endogenous triggering of biostimulants [60]. At least 60 essential oil constituents were identified in elderberry (*Sambucus ebulus* L.), with some of the components significantly increasing in content under different exogenous application of auxin (IAA and NAA) [133]. Treatment with CK increased the total oil yield of peppermint plants by 40% [77].

Jasmonates are directly involved in the mevalonic acid pathway, through the enzyme mevalonate-5-diphosphate carboxylase, which directly affects the biosynthesis of linalool [134]. Linalool levels decreases with increases in excess endogenous MeJA, which mostly accumulated following simulated hail damage [122]. In other studies, [135, 136] demonstrated that applications over 18 μM (MeJA) may significantly affect the accumulation of linalool content.

Biosynthesis of isomenthone occurs late in leaf development, when mature oil gland cells are in the post-secretory phase [137]. At this late stage of leaf development, ABA is abundant in the epidermal cells, where it is involved in the abscission process [138]. In the study conducted by [122], the daily application of ABA led to an excess of endogenous ABA content, causing toxicity affecting the biosynthesis of isomenthone. Similarly, [60] stated that the chemical composition of essential oil plants could be influenced by exogenous applications of ABA and methyl jasmonate.

In plants, geraniol and ABA biosynthesis share a similar pathway [41]: the ABA biosynthetic pathway starts from oxidative cleavage of the epoxy-carotenoids and 9-*cis* violaxanthin, where xanthoxin is converted to abscisic acid. The process is then followed by sequential production of farnesyl pyrophosphate, phytoene, carotene, lycopene, and geranyl pyrophosphate [41]. Abscisic acid and geraniol biosynthesis occur at this stage through the isopentenyl diphosphate source [139]. Croteau and

Purkett [140] reported that geranyl pyrophosphate synthase activity is localized in the leaf epidermal glands of sage (*Salvia officinalis* L.), where monoterpenes are biosynthesized, suggesting that geranyl pyrophosphate synthase supplies the C10 precursor for the production of monoterpenes. In this case, any repeated application of ABA can reach a toxic level in the epidermal cells as described by [122, 139], causing disruptions of geraniol biosynthesis through the cytosolic mevalonate pathway.

Geraniol usually undergoes biotransformation into other terpenoids in aromatic plants, which influences the quality of the essential oil [141]. Geranyl formate, geranyl butyrate, geranyl tiglate, and geranyl acetate are some of the acyclic monoterpenes derived from geraniol, which are regarded as geraniol esters [142]. It has been noted in the study conducted by [122] that any prolonged application of a low concentration of MeJA increases the content of geranyl tiglate. This was supported by [143], who described that jasmonates are upregulated by wounding stress and are directly involved in the biosynthesis of these terpenes. In addition, the accumulation of geranyl tiglate can be attributed to simulated hail damage as described by [122], followed by the subsequent daily application of methyl jasmonate. The accumulation of geranyl formate is attributed to the biosynthesis of geraniol, and the effects of subsequent daily use of methyl jasmonate [142].

Plants respond to defoliation and wounding through the induction of phenylpropanoids metabolism to accumulate phenolic compounds [144]. Phenolics provide cytotoxic effects, as well as the building blocks for polyphenolic-based cell wall modifications. Polyphenolic-based cell wall modifications assist with organ regeneration in defoliated or wounded plants [118, 144]. The level of endogenous phenolics in refoliated plants influences plant growth and development, clearly indicating the relation of the leaf refoliation with the leaf ontogeny [97]. The age of the plant is associated with the level of the total phenolics. On wampee (*Clausena lansium* Lour.), [145] reported high total leaf phenolics in early growth stages compared to the later growth stages. A high phenolic content was also observed in the early stages of leaf development of yerba-mate (*Ilex paraguariesnsis* A.St Hil.), which affected the development and quality of these plants [146]. High phenolics level in young leaves may be associated with defense and refoliation following defoliation [147].

Previous studies have demonstrated that defoliation, wounding and exogenous application of biostimulants on plants may effectively stimulate vegetative growth, improve nutrient acquisition, and increases the antioxidant capacity of plant tissue [69]. Aspects of phenolics accumulation are driven by upregulation of endogenous phytohormones. However, this occurs through crosstalk networks between these phytohormones, for which some commercially available [148]. A combination of spermine, methyl jasmonate, and epibrassinolide was found to induce secondary metabolites, including phenolics, in sweet basil (*Ocimum basilicum* L.) [149]. According to [150], maintenance of minerals is a prerequisite for providing co-factors for many enzymes of the phenolic pathway. The accumulation of caffeic acid, caffeoyl tartaric acid, p-Coumaroyl tyrosine and protocatechuic acid O-hexoside are a typical plant response to defoliation, including wounding stress [118, 151].

Biostimulants stimulate plant growth and terpene biosynthesis in a large number of aromatic plant species, which result in beneficial changes in terpene accumulation [152]. Poyh and Ono [153] recorded higher essential oil content for sage (*S. officinalis* L.) treated with 100 mg/L gibberellic acid. Fraternale et al. [154] reported a higher essential oil yield of Spanish marjoram (*Thimus mastichina* L.) using a medium culture with CK as low as 0.1 mg/L. In the leaves of Spanish marjoram treated with CK, there were a greater number of glandular trichomes recorded at the later leaf developmental stage, which could be ascribed to increased essential oil yield. Foliar application of a biostilumat-28-homobrassinolide (10^{-6} M) also enhanced the essential oil yield of mint (*M. arvensis* L.) [155].

4. Conclusions

Essential oil plants are mostly grown for essential oil production; therefore, the aerial herbage material is a crucial component of these crops. Environmental stress factors, such as hail damage, can cause significant damage to these plants, reducing this valuable material, and directly affecting the essential oil yield and quality [72]. Natural and synthetic biostimulants have been extensively investigated on plant growth and development, and also on the recovery of plants following stress [60]. Based on [60], it was hypothesized that the application of biostimulants will recover the herbage yield and improve the biosynthesis of essential oil plants subjected to simulated hail damage. This chapter has detailed the potential recovery response of plants to hail damage, the response of plants treated with synthetic biostimulants, and the response of plants treated with natural biostimulants extracted from plants. Therefore, it is evident that the use of natural or synthetic biostimulants, as an alternative mitigation strategy against hail damage, might help in the recovery and improve essential oil plant yield following hail damage. However, future studies should explore;

1. the extents of hail damage on different essential oil plants;
2. determining the effects of root-applied synthetic cytokinin on the recovery of essential oil yield attributes, and the essential oil yield and quality of hail-damaged essential oil plants.
3. determining the effects of cytokinin and the auxin ratio on hail-damaged essential oil plants, *in vivo* study.
4. identifying the effects of synthetic biostimulants as a contaminant of essential oil quality: A perfumery industry study.
5. evaluate the use of combined plant growth regulator as a potential recovery mechanism of the herbage yield, and the essential oil yield and quality of simulated hail-damaged essential oil plants, and,
6. evaluate the use of abscisic acid and methyl jasmonate as a potential mitigating mechanism on simulated hail-damaged rose geranium plants.

Conflict of interest

The authors declare no conflict of interest.

Author details

Zenzile Peter Khetsha[1*], Moosa Mahmood Sedibe[1], Rudolph Johannes Pretorius[1], Phoka Caiphus Rathebe[2] and Karabelo Moloantoa[3]

1 Central University of Technology, Free State, Bloemfontein, South Africa

2 University of Johannesburg, Johannesburg, South Africa

3 University of the Free State, Bloemfontein, South Africa

*Address all correspondence to: zkhetsha@cut.ac.za

References

[1] Halsnæs K, Larsen MAD, Kaspersen PS. Climate change risks for severe storms in developing countries in the context of poverty and inequality in Cambodia. National Hazards. 2018;**94**:261-278

[2] Gobbo S, Ghiraldini A, Dramis A, Dal Ferro N, Morari F. Estimation of hail damage using crop models and remote sensing. Remote Sensing. 2021;**13**(14):2655

[3] Khetsha ZP, Sedibe MM, Pretorius RJ. Effects of abscisic acid and methyl jasmonate on the recovery of hail damaged rose geranium (*Pelargonium graveolens*) plants. Acta Horticulturae. 2020;**1287**:31-40

[4] Kundu S, Gantait S. Abscisic acid signal crosstalk during abiotic stress response. Plant Gene. 2017;**11**:61-69

[5] Jamli JB. Physical analysis of hail fall risk in Iran and the consequent damages on agricultural crops. Atmospheric and Climate Sciences. 2014;**4**:919-930

[6] Bal SK, Saha S, Fand BB, Singh NP, Rane J, Minhas PS. Hailstorms: Causes, Damage and Post-Hail Management in Agriculture. India: National Institute of Abiotic Stress Management Technical Bulletin-5; 2014

[7] Weiss EA. Essential Oil Crops. United Kingdom: CAB International; 1997

[8] Khetsha ZP, Sedibe MM, Pretorius RJ, Van Der Watt E. Plant growth regulators and simulated hail-damage improve rose-geranium (*Pelargonium graveolens* L'Hér.) leaf mineral status and phenolic composition. Applied Ecology and Environmental Research. 2021;**19**(4):3083-3095

[9] Banchio E, Valladares G, Zygadlo J, Bogino PC, Rinaudi LV, Giordano W. Changes in composition of essential oils and volatile emissions of *Minthostachys mollis*, induced by leaf punctures of *Liriomyza Huidobrensis*. Biochemical Systematics and Ecology. 2007;**35**: 68-74

[10] Troncoso C, Becerra J, Bittner M, Perez C, Saez K, Sanchez-Olate M, et al. Chemical defense responses in *Eucalyptus globulus* (Labill) plants. Journal of the Chilean Chemical Society. 2011;**56**:768-770

[11] Potter LE, Brotzge JA, Erickson SA. Warning the public about hail: Determining the potential for short-term damage mitigation. In: Preprints, 5th Symposium. Policy and Socio-economic Research. Atlanta, United States of America: American Meteorological Society; 2010

[12] Marè FA. Evaluating crop insurance as production risk management strategy [MSc thesis]. South Africa: University of the Free State; 2014

[13] Henson RA, Heichel GH. Nitrogen accumulation and partitioning in hail-damaged soybeans. Journal of Plant Nutrition. 1986;**9**:1453-1468

[14] Araya HT, Soundy P, Steyn JM, Teubes C, Learmonth RA, Nare M. Response of herbage yield, essential oil yield and composition of South African rose-scented geranium (*Pelargonium* spp.) to conventional and organic nitrogen. Journal of Essential Oil Research. 2006;**18**:111-115

[15] Sedibe MM. Yield and quality response of hydroponically grown rose geranium (*Pelargonium* spp.) to changes in the nutrient solution and shading [PhD thesis]. South Africa: University of the Free State; 2012

[16] Yu L, Liu H, Shao X, Yu F, Wei Y, Ni Z, et al. Effects of hot air and methyl jasmonate treatment on the metabolism of soluble sugars in peach fruit during cold storage. Postharvest Biology and Technology. 2016; **113**:8-16

[17] Han Y, Ishibashi S, Iglesias-Gonzalez J, Chen Y, Love NR, Amaya E. Ca^{2+}-induced mitochondrial ROS regulate the early embryonic cell cycle. Cell Reports. 2018; **22**:218-231

[18] Ju Y, Min L, Zhao H, Meng J, Fang Y. Effect of exogenous abscisic acid and methyl jasmonate on anthocyanin composition, fatty acids, and volatile compounds of cabernet sauvignon (*Vitis vinifera* L.) grape berries. Molecules. 2016;**21**:1354

[19] Nemhauser JL, Hong FX, Chory J. Different plant hormones regulate similar processes through largely non-overlapping transcriptional responses. Cell. 2006;**126**:467-475

[20] Giannakoula A, Ilias IF, Maksimović JD. The effects of plant growth regulators on growth, yield and phenolic profile of lentil plants. Journal of Food Composition and Analysis. 2012;**28**:48-53

[21] Anderson JP. Antagonistic interaction between abscisic acid and jasmonate-ethylene signalling pathways modulates defence gene expression and disease resistance in Arabidopsis. The Plant Cell Online. 2004;**16**:3460-3479

[22] Drábková LZ, Dobrev PI, Motyka V. Phytohormone profiling across the bryophytes. PLoS One. 2015;**10**:5

[23] Mills E. Insurance in a climate of change. Science. 2005;**309**:1040-1044

[24] Fernandes GW, Oki Y, Sá D, Sales MCE, Quintino MN, Freitas AV, et al. Hailstorm impact across plant taxa: Leaf fall in a mountain environment. Neotropical Biology and Conservation. 2012;**7**:8-15

[25] Steiner JT. Can we reduce the hail problem? Weather and Climate. 1988;**8**:23-32

[26] Schubert T. Hail damage to plants. Plant Pathology. 1991;**347**:1-2

[27] Powell CL. Radar characteristics of hailstorms in South Africa the summer rainfall regions of South Africa frequently experience thunderstorms between September and April. In: 29th Annual Conference of South African Society for Atmospheric Sciences SASAS 2013; 26-27 September 2013; Durban, South Africa. 2013

[28] Łukaszuk E, Ciereszko I. Plant responses to wounding stress. Biological Diversity—from Cell to Ecosystem. 2012;**2012**:73-85

[29] Changnon SA. Data and approaches for determining hail risk in the contiguous United States. Journal of Applied Meteorology. 1999;**38**:1730-1739

[30] Klein RN, Shapiro CA. Evaluating Hail Damage to Corn. United States of America: BRANDT Agronomy Bulletin; 2011

[31] Suttle JC, Lulai E, Huckle LL, Neubauer JD. Wounding of potato tubers induces increases in ABA biosynthesis

and catabolism and alters expression of ABA metabolic genes. Journal of Plant Physiology. 2013;**170**:560-566

[32] Smékalová V, Doskočilova A, Komis G, Šamaj J. Crosstalk between secondary messengers, hormones and MAPK modules during osmotic stress signalling in plants. Biotechnology Advances. 2014;**32**:2-11

[33] Sanchez-Serrano J, Schmidt R, Schell J, Willmitzer L. Nucleotide sequence of proteinase inhibitor II encoding cDNA of potato (*Solanum tuberosum*) and its mode of expression. Molecular and General Genetics. 1986;**203**:15-20

[34] Christopher ME, Miranda M, Major IT, Constabel CP. Gene expression profiling of systemically wound-induced defenses in hybrid poplar. Planta. 2004;**219**:936-947

[35] Kang HM, Saltveit ME. Wound-induced increases in phenolic content of fresh-cut lettuce is reduced by a short immersion in aqueous hypertonic solutions. Postharvest Biology and Technology. 2003;**29**:271-277

[36] León J, Rojo E, José J, Sánchez-Serrano S. Wound signalling in plants. Journal of Experimental Botany. 2001;**52**:1-9

[37] Torres N, Goicoechea N, Antolín MC. Influence of irrigation strategy and mycorrhizal inoculation on fruit quality in different clones of Tempranillo grown under elevated temperatures. Agricultural Water Management. 2018;**202**:285-298

[38] Jindo K, Canellas LP, Albacete A, Figueiredo dos Santos L, Frinhani Rocha RL, Carvalho Baia D, et al. Interaction between humic substances and plant hormones for phosphorous acquisition. Agronomy. 2020;**10**(5):640

[39] Mano Y, Nemoto K. The pathway of auxin biosynthesis in plants. Journal of Experimental Botany Advance Access. 2012;**63**:2853-2872

[40] Gaspar T, Kevers C, Penel C, Greppin H, Reid DM, Thorpe TA. Plant hormones and plant growth regulators in plant tissue culture. In Vivo Cell Development Biology Plant. 1996;**32**:272-289

[41] Olds CL, Glennon EK, Luckhart S. Abscisic acid: New perspectives on an ancient universal stress signaling molecule. Microbes and Infection. 2018;**20**:484-492

[42] Saito S, Okamoto M, Shinoda M, Kushiro T, Koshiba T, Kamiya Y, et al. A plant growth retardant, uniconazole, is a potent inhibitor of ABA catabolism in Arabidopsis. Bioscience, Biotechnology and Biochemistry. 2006;**70**:1731-1739

[43] Tardieu F, Parent B, Simonneau T. Control of leaf growth by abscisic acid: Hydraulic or non-hydraulic processes? Plant, Cell & Environment. 2009;**33**:636-647

[44] Punpee P. Conditional up-regulation of cytokinin status increases growth and survival of sugarcane in water-limited onditions [PhD thesis]. Australia: University of Queensland; 2015

[45] Dammann C, Rojo E, Sanchez-Serrano JJ. Abscisic acid and jasmonic acid activate wound inducible genes in potato through separate, organ-specific signal transduction pathways. The Plant Journal. 1997;**11**:773-782

[46] Seo M, Koshiba T. Transport of ABA from the site of biosynthesis to the site of action. Journal of Plant Research. 2011;**124**:501-507

[47] Tillberg E. Effect of light on abscisic acid content in photosensitive Scots pine (*Pinus sylvestris* L.) seed. Plant Growth Regulation. 1992;**11**:147-152

[48] Kong Z, Zhao D. The inhibiting effect of abscisic acid on fragrance of Kam sweet rice. Journal of Food and Nutrition Research. 2014;**2**:148-154

[49] Sansberro PA, Mrogiski LA, Bottini R. Foliar sprays with ABA promote growth of *Ilex paraguariensis* by alleviating diurnal water stress. Plant Growth Regulation. 2004;**42**:105-111

[50] Peña-Cortés H, Sanchez-Serrano JJ, Mertens R, Willmitzer L, Salome P. Abscisic acid is involved in the wound-induced expression of the proteinase inhibitor II gene in potato and tomato. Proceedings of the National Academy of Sciences, United States of America. 1989;**86**:9851-9855

[51] Herde O, Peña-Cortés H, Wasternack C, Willmitzer L, Fisahn J. Electricsignallingand*pin2*geneexpression on different abiotic stimuli depend on a distinct threshold level of endogenous abscisic acid in several abscisic acid-deficient tomato mutants. Plant Physiology. 1999;**119**:213-218

[52] Wang T, Li C, Wu Z, Jia Y, Wang H, Sun S, et al. Abscisic acid regulates auxin homeostasis in rice root tips to promote root hair elongation. Frontiers of Plant Science. 2018;**8**:1121

[53] Dar TA, Uddin M, Khan MMA, Hakeem KR, Jaleel H. Jasmonates counter plant stress: A review. Environmental and Experimental Botany. 2015;**115**:49-57

[54] Piotrowska A, Bajguz A. Conjugates of abscisic acid, brassinosteroids, ethylene, gibberellins, and jasmonates. Phytochemistry. 2011;**72**:2097-2112

[55] Huang H, Liu B, Liu L, Song S. Jasmonate action in plant growth and development. Journal of Experimental Botany. 2017;**68**:1349-1359

[56] Turner JG, Ellis C, Devoto A. The jasmonate signal pathway. Plant Cell. 2002;**14**:S153-S164

[57] Piotrowska A, Bajguz A, Godlewska-Żyłkiewicz B, Czerpak R, Kamińska M. Jasmonic acid as modulator of lead toxicity in aquatic plant *Wolffia arrhiza* (Lemnaceae). Environmental and Experimental Botany. 2009;**66**:507-513

[58] Guo Q, Major IT, Howe GA. Resolution of growth-defense conflict: Mechanistic insights from jasmonate signalling. Current Opinion in Plant Biology. 2018;**44**:E10768-E10777

[59] Koo AJK, Howe GA. The wound hormone jasmonate. Photochemistry. 2009;**70**:1571-1580

[60] Prins CL, Viera IJC, Freitas SP. Growth regulators and essential oil production. Brazilian Society of Plant Physiology. 2010;**22**:91-102

[61] Xue R, Zhang B. Increased endogenous methyl jasmonate altered leaf and root development in transgenic soybean plants. Journal of Genetics and Genomics. 2007;**34**:339-346

[62] Anderson JM. Jasmonic acid-dependent increase in the level of specific polypeptides in soybean suspension cultures and seedlings. Journal of Plant Growth Regulation. 1988;**7**:203-211

[63] Kim HJ, Chen F, Wang X, Rajapaske NC. Effect of methyl jasmonate on secondary metabolites of sweet basil (*Ocimum basilicum* L.). Journal of Agricultural and Food Chemistry. 2006;**54**:2327-2332

[64] Degenhardt DC, Lincoln DE. Volatile emissions from an odorous plant in response to herbivory and methyl jasmonate exposure. Journal of Chemical Ecology. 2006;**32**:725-743

[65] Thaler JS, Farag MA, Paré PW, Dicke M. Jasmonate-deficient plants have reduced direct and indirect defences against herbivores. Ecology Letters. 2002;**5**:764-774

[66] Loivamäki M, Holopainen JK, Nerg AM. Chemical changes induced by methyl jasmonate in oilseed rape grown in the laboratory and in the field. Journal of Agricultural and Food Chemistry. 2004;**52**:7607-7613

[67] Konan YK, Kouassi KM, Kouakou KL, Koffi E, Kouassi KN, Sekou D, et al. Effect of methyl jasmonate on phytoalexins biosynthesis and induced disease resistance to *Fusarium oxysporum* f. sp. Vasinfectum in cotton (*Gossypium hirsutum* L.). International Journal of Agronomy. 2014: Article ID 806439

[68] Jiang Y, Ye J, Li S, Niinemets Ü. Methyl jasmonate-induced emission of biogenic volatiles is biphasic in cucumber: A high-resolution analysis of dose dependence. Journal of Experimental Botany. 2017;**68**:4679-4694

[69] Parađiković N, Teklić T, Zeljković S, Lisjak M, Špoljarević M. Biostimulants research in some horticultural plant species-A review. Food and Energy Security. 2019;**8**:e00162

[70] Taiz L, Zeiger E. Plant Physiology. Sunderland, MA: Sinauer; 2002

[71] Cato SC, Macedo WR, Peres LEP, de Castro PR. Synergism among auxins, gibberellins and cytokinins in tomato cv. micro-tom. Horticultura Brasileira. 2013;**31**:549-553

[72] Weiss D, Ori N. Mechanisms of cross talk between gibberellin and other hormones. Plant Pathology. 2007;**144**:1240-1246

[73] Sharma S, Sharma A, Kaur M. Extraction and evaluation of gibberellic acid from *Pseudomonas* spp. plant growth-promoting rhizobacteria. Journal of Pharmacognosy and Phytochemistry. 2018;**7**:2790-2795

[74] Kieber JJ. Tribute to Folke Skoog: Recent advances in our understanding of cytokinin biology. Journal of Plant Growth Regulation. 2002;**21**:1-2

[75] Koprna R, De Diego N, Dundálková L, Spíchal L. Use of cytokinins as agrochemicals. Bioorganic and Medicinal Chemistry. 2016;**24**:484-492

[76] Kudoyarova GR, Dodd IC, Veselov DS, Rothwell SA, Veselov SY. Common and specific responses to availability of mineral nutrients and water. Journal of Experimental Botany. 2015;**66**:2133-2144

[77] Santoro MV, Nievas F, Zygadlo J, Giordano W, Banchio E. Effects of growth regulators on biomass and the production of secondary metabolites in peppermint (*Mentha piperita*) micropropagated in-vitro. American Journal of Plant Sciences. 2013;**4**:49-55

[78] Egamberdieva D, Wirth SJ, Alqarawi AA, Abd_Allah EF, Hashem A. Phytohormones and beneficial microbes: Essential components for plants to balance stress and fitness. Frontiers in Microbiology. 2017;**8**:2104

[79] Hedden P, Sponsel VA. Century of gibberellin research. Journal of Plant Growth Regulation. 2015;**34**:740-760

[80] Van Der Watt E. Bio-stimulatory properties of a *Lupinus albus* L. seed suspension [PhD thesis]. South Africa: University of the Free State; 2005

[81] Hooykaas PJ, Hall MA, Libbenga KR. Regulation of gibberellin biosynthesis. Biochemistry and Molecular Biology of Plant Hormones. 1999;**13**:161

[82] Sakamoto T, Kamiya N, Ueguchi-Tanaka M, Iwahori S, Matsuoka M. KNOX homeodomain protein directly suppresses the expression of a gibberellin biosynthetic gene in the tobacco shoot apical meristem. Genes Development. 2001;**15**:581-590

[83] Leite VM, Rosolem CA, Rodrigues JD. Gibberellin and cytokinin effects on soybean growth. Scientia Agricola. 2003;**60**:537-541

[84] Greenboim-Wainberg Y, Mayhom I, Borochov R, Alvarez J, Olszewski N, Ori N, et al. Crosstalk between gibberellic and cytokinin: The Arabidopsis GA response inhibitor SPINDLY plays a positive role in cytokinin signalling. Plant Cell. 2005;**17**:92-102

[85] Su YH, Liu YB, Zhang XS. Auxin–cytokinin interaction regulates meristem development. Molecular Plant. 2011;**4**:616-625

[86] Moubayidin L, Perilli S, Ioio RD, Di Mambro R, Costantino P, Sabatini S. The rate of cell differentiation controls the Arabidopsis root meristem growth phase. Current Biology. 2010;**20**:1138-1143

[87] Verma P, Sen NL. The impact of plant growth regulators on growth and biochemical constituents of coriander (*Coriandrum sativum* L.). Journal of Herbs, Spices and Medicinal Plants. 2008;**14**:144-153

[88] Schoene G, Yeager T. Effects of benzyladenine applied alone or in combination with gibberellic acid in sweet viburnum. SNA Research Conference. 2005;**50**:114-122

[89] Barnes HW. Gibberellins and cytokinins: A review. Acta Horticulturae. 2013;**1055**:323-336

[90] Hüster T. Growth and yield response of selected crops to treatment with ComCat [PhD thesis]. South Africa: University of the Free State; 2011

[91] Li X, Zhang L, Ahammed GJ, Li ZX, Wei JP, Shen C, et al. Nitric oxide mediates brassinosteroid-induced flavonoid biosynthesis in *Camellia sinensis* L. Journal of Plant Physiology. 2017;**214**:145-151

[92] Tang J, Han Z, Chai J. Q&A: What are brassinosteroids and how do they act in plants? BMC Biology. 2016;**14**:113

[93] Sirhindi G, Kaur H, Bhardwaj R, Kaur NP, Sharma P. Therno protective role of 28-homobrassinolide on the growth and antioxidant enzymes activities in the seedling of *Brassica juncea* L. Physiology and Molecular Biology of Plants. 2014;**5**:48074-48079

[94] Rao SS, Vardhini BV, Sujatha E, Anuradha S. Brassinosteroids—A new class of phytohormones. Current Science. 2002;**25**:1239-1245

[95] Yin W, Dong N, Niu M, Zhang X, Li L, Liu J, et al. Brassinosteroid-regulated plant growth and development and gene expression in soybean. The Crop Journal. 2019;**7**:411-418

[96] Alyemeni M, Al-Quwaiz SM. Effect of 28-homobrassinolide on the performance of sensitive and resistant varieties of *Vigna Radiata*. Saudi Journal of Biological Sciences. 2016;**23**:698-705

[97] Kulaeva ON, Burkhanova EA, Fedina AB, Khokhlova VA, Bokebayeva GA, Vorbrodt HM, et al. Effect of brassinosteroids on protein synthesis and plant-cell ultrastructure under stress conditions. ACS Symposium Series. 1991;**474**:141-155

[98] Maia JGS, Andrade EHA, Couto HA, da Silva ACM, Marx F, Henke C. Plant sources of Amazon rosewood oil. Quimica Nova Chemistry in Brazil. 2007;**30**:1906-1910

[99] Senthil A, Pathmanaban G, Srinivasan PS. Effect of bioregulators on some physiological and biochemical parameters of soybean (*Glycine max* L.). Legume Research—An International Journal. 2003;**26**:54-56

[100] Vardhini BA, Anuradha S, Sujatha E, Rao SSR. Role of brassinosteroids in alleviating various abiotic and biotic stresses–A review. Plant Stress. 2010;**4**:55-61

[101] Sezgin M, Kahya M. Phytohormones. Bitlis Eren University Journal of Science and Technology. 2018;**8**:35-39

[102] Bulgari R, Cocetta G, Trivellini A, Vernieri P, Ferrante A. Biostimulants and crop responses: A review. Biological Agriculture and Horticulture. 2015;**31**:1-17

[103] Michalski T. Possibilities of Maize Production Increase Using Non-conventional Technologies: Biostimulators in Modern Agriculture, Part I. General aspects. Warsaw: Editorial House Wieś Jutra; 2008

[104] Yakhin OI, Lubyanov AA, Yakhin IA, Vakhitov VA, Ibragimov RI, Yumaguzhin MS, et al. Metabolic changes in wheat (*Triticum aestivum* L.) plants under action of bioregulator stifun. Applied Biochemistry and Microbiology. 2011;**47**:621

[105] Carillo P, Colla G, El-Nakhel C, Bonini P, D'Amelia L, Dell'Aversana E, et al. Biostimulant application with a tropical plant extract enhances Corchorus olitorius adaptation to sub-optimal nutrient regimens by improving physiological parameters. Agronomy. 2019;**9**:249

[106] Machado VPO, Pacheco AC, Carvalho MEA. Effect of biostimulant application on production and flavonoid content of marigold (*Calendula officinalis* L.). Revista Ceres. 2014;**61**:983-988

[107] El-Khateeb MA, El-Attar AB, Nour RM. Application of plant biostimulants to improve the biological responses and essential oil production of marjoram (*Majorana hortensis*, Moench) plants. Middle East Journal of Agriculture Research. 2017;**6**:928-941

[108] Sarraf M, Kataria S, Taimourya H, Santos LO, Menegatti RD, Jain M, et al. Magnetic field (MF) applications in plants: An overview. Plants (Basel, Switzerland). 2020;**9**(9):1139

[109] Konefał-Janocha M, Banaś-Ząbczyk A, Bester M, Bocak D, Budzik S, Górny S, et al. The effect of stationary and variable electromagnetic fields on the germination and early growth of radish (*Raphanus sativus*). Polish Journal of Environmental Studies. 2018;**28**(2):709-715

[110] Afzal I, Saleem S, Skalicky M, Javed T, Bakhtavar MA, Kamran M, et al. Magnetic field treatments improves sunflower yield by inducing physiological and biochemical modulations in seeds. Molecules. 2021;**26**(7):20-22

[111] Vashisth A, Joshi DK. Growth characteristics of maize seeds exposed

to magnetic field. Bioelectromagnetics. 2017;**38**(2):151-157

[112] Kırdar E, Yücedağ C, Balaban B. The effects of magnetic field on germination of seeds and growth of seedlings of stone pine. Journal of Forests. 2016;**3**(1):1-6

[113] Vian A, Davies E, Gendraud M, Bonnet P. Plant responses to high frequency electromagnetic fields. BioMed Research International. 2016;**2**:1-13

[114] Podleśna A, Bojarszczuk J, Podleśny J. Effect of pre-sowing magnetic field treatment on some biochemical and physiological processes in faba bean (*Vicia faba* L. spp. Minor). Journal of Plant Growth Regulation. 2019;**38**(3):1153-1160

[115] Govindaraj M, Masilamani P, Albert VA, Bhaskaran M. Effect of physical seed treatment on yield and quality of crops: A review. Agricultural Reviews. 2017;**38**(1):1-14

[116] Vashisth A, Meena N, Krishnan P. Magnetic field affects growth and yield of sunflower under different moisture stress conditions. Bioelectromagnetics. 2021;**42**(6):473-483

[117] Ikeuchi M, Iwase A, Rymen B, Lambolez A, Kojima M, Takebayashi Y, et al. Wounding triggers callus formation via dynamic hormonal and transcriptional changes. Plant Physiology. 2017;**175**:1158-1174

[118] Nanda AK, Melnyk CW. The role of plant hormone during grafting. Semi-in-vivo Developmental Biology. 2018;**131**:49-58

[119] Oosthuizen L, Coetzee J. Morphogenesis of trichomes of *Perlagonium scabrum*. South African Journal of Botany. 1984;**2**:305-310

[120] Ma D, Hu Y, Yang C, Liu B, Fang L, Wan Q, et al. Genetic basis for glandular trichome formation in cotton. Nature Communications. 2016;**7**:1-9

[121] Addicott FF, Lynch RS. Physiology of abscission. Annual Review of Plant Physiology. 1955;**6**:211-238

[122] Khetsha ZP. Using agricultural plant growth regulators as a mitigating strategy for hail-damaged rose geranium (*Pelargonium graveolens* L'Her) [PhD thesis]. South Africa: Central University of Technology, Free State; 2020

[123] Kalve S, De Vos D, Beemster GT. Leaf development: A cellular perspective. Frontiers in Plant Science. 2014;**5**:362

[124] Zhou Z, An L, Sun L, Zhu S, Xi W, Broun P, et al. Zinc Finger Protein5 is required for the control of trichome initiation by acting upstream of Zinc Finger Protein8 in Arabidopsis. Plant Physiology. 2011;**157**:673-682

[125] Zhou Z, Sun L, Zhao Y, An L, Yan A, Meng X, et al. Zinc Finger Protein 6 (ZFP6) regulates trichome initiation by integrating gibberellin and cytokinin signaling in *Arabidopsis thaliana*. New Phytologist. 2013;**198**:699-708

[126] Xue S, Dong M, Xingwang L, Xu S, Pang J, Zhang W, et al. Classification of fruit trichomes in cucumber and effects of plant hormones on type II fruit trichome development. Planta. 2019;**249**:407-416

[127] Campos ML, de Almeida M, Rossi ML, Martinelli AP, Junior CGL, Figueira A, et al. Brassinosteroids interact negatively with jasmonates in the formation of anti-herbivory traits in tomato. Journal of Experimental Botany. 2009;**60**:4347-4361

[128] Fambrini M, Pugliesi C. The dynamic genetic-hormonal regulatory network controlling the trichome development in leaves. Plants. 2019;**8**:253

[129] Pattanaik S, Patra B, Singh SK, Yuan L. An overview of the gene regulatory network controlling trichome development in the model plants, Arabidopsis. Frontier in Plant Science. 2014;5:259

[130] Stobbe H, Schmitt U, Eckstein D, Dujesiefken D. Developmental stages and fine structure of surface callus formed after debarking of living lime trees (*Tilia* sp.). Annals of Botany. 2002;**89**:773-782

[131] Brilli F, Ruuskanen TM, Schinitzhofer R, Müller M, Breitenlechner M, Bittner V, et al. Detection of plant volatiles after leaf wounding and darkening by proton transfer reaction "time-of-flight" mass spectrometry (PTR-TOF). PLoS One. 2011;**6**:e20419

[132] Rajeswara-Rao BR. Biomass yield, essential oil yield and essential oil composition of rose-scented geranium (*Pelargonium* spp.) as influenced by row spacings and intercropping with corn mint (*Mentha arvensis* L.f. Piperascens Malinv. Ex Holmes). Industrial Crops and Products. 2000;**16**:133-144

[133] Feibakhsh A, Pazoki H, Vahid Mohammadrezaei V, Ebrahimzadeh MA. Effect of phytohormones on the composition of *Sambucus ebulus* leaf essential oil. Tropical Journal of Pharmaceutical Research. 2014;**13**:581-586

[134] Bruno M, Llardi V, Lupidi G, Quassinti L, Bramucci M, Fiorini D, et al. The non-volatile and volatile metabolites of *Prangos ferulacea* and their biological properties. Planta Medica. 2019;**85**:815-824

[135] Rodriguez-Saona C, Crafts-Brandner SJ, Williams L, Pare P. *Lygus hesperus* feeding and salivary gland extracts induce volatile emissions in plants. Journal of Chemical Ecology. 2002;**28**:1733-1747

[136] Van Schie CN, Haring MA, Schuurink RC. Tomato linalool synthase is induced in trichomes by jasmonic acid. Plant Molecular Biology. 2007;**64**:251-263

[137] Davis EM, Ringer KL, McConkey ME, Croteau R. Monoterpene metabolism, cloning, expression, and characterization of menthone reductases from peppermint. Plant Physiology. 2005;**137**:873-881

[138] Wilmowicz E, Frankowski K, Kućko A, Świdziński M, de Dios AJ, Nowakowska A, et al. The influence of abscisic acid on the ethylene biosynthesis pathway in the functioning of the flower abscission zone in *Lupinus luteus*. Journal of Plant Physiology. 2016;**206**:49-58

[139] Burlat V, Oudin A, Courtois M, Rideau M, St-Pierre B. Co-expression of three MEP pathway genes and geraniol 10-hydroxylase in internal phloem parenchyma of *Catharanthus roseus* implicates multicellular translocation of intermediates during the biosynthesis of monoterpene indole alkaloids and isoprenoid-derived primary metabolites. Plant Journal. 2004;**38**:131-141

[140] Croteau R, Purkett PT. Geranyl pyrophosphate synthase: Characterization of the enzyme and evidence that this chain-length specific prenyltransferase is associated with monoterpene biosynthesis in sage (*Salvia officinalis*). Archives of Biochemistry and Biophysics. 1989;**271**:524-535

[141] Demyttenaere JC, del Carmen HM, De Kimpe N. Biotransformation of geraniol, nerol and citral by sporulated

surface cultures of *Aspergillus niger* and *Penicillium* sp. Phytochemistry. 2000;**55**:363-373

[142] Liu W, Xu X, Zhang R, Cheng T, Cao Y, Li X, et al. Engineering *Escherichia coli* for high-yield geraniol production with biotransformation of geranyl acetate to geraniol under fed-batch culture. Biotechnology for Biofuels. 2016;**9**:1-8

[143] Zhang G, Zhao F, Chen L, Pan Y, Sun L, Bao N, et al. Jasmonate-mediated wound signalling promotes plant regeneration. Nature Plants. 2019;**5**:491-497

[144] Bernards MA, Båstrup-Spohr L. Phenylpropanoid Metabolism Induced by Wounding and Insect Herbivory. Dordrecht: Springer; 2008

[145] Chang X, Lu Y, Lin Z, Qiu J, Guo X, Pan J, et al. Impact of leaf development stages on polyphenolics profile and antioxidant activity in *Clausena lansium* (Lour.) skeels. Hindawi. BioMed Research International. 2018;**2018**:7093691

[146] Blum-Silva CH, Chavesa VC, Schenkela EP, Coelhob GC, Reginattoa FH. The influence of leaf age on methylxanthines, total phenolic content, and free radical scavenging capacity of *Ilex paraguariensis* aqueous extracts. Revista Brasileira de Farmacognosia. 2015;**25**:1-6

[147] Yoshioka T, Inokuchi T, Fujioka S, Kimura Y. Phenolic compounds and flavonoids as plant growth regulators from fruit and leaf of *Vitex rotundifolia*. Zeitschrift für Naturforschung C Journal of Biosciences. 2004;**59**:509-514

[148] Wani SH, Kumar V, Shriram V, Kumar SS. Phytohormones and their metabolic engineering for abiotic stress tolerance in crop plants. The Crop Journal. 2016;**4**:162-176

[149] Koca N, Karaman Ş. The effects of plant growth regulators and L-phenylalanine on phenolic compounds of sweet basil. Food Chemistry. 2014;**166**:515-521

[150] Treutter D. Managing phenol contents in crop plants by phytochemical farming and breeding-visions and constraints. International Journal of Molecular Sciences. 2010;**11**:807-857

[151] Bryant JP, Reichardt PB, Clausen TP, Werner RA. Effects of mineral nutrition on delayed inducible resistance in Alaska paper birch. Ecology. 1993;**74**:2072-2084

[152] Farooqi AHA, Shukla A. Utilization of plant growth regulators in aromatic plant production. Chromatography. 1990;**12**:152-157

[153] Poyh JA, Ono EO. Performance of the essential oil of *Salvia officinalis* L. under the action of plant regulators. Acta Scientiarum Biological Sciences. 2006;**28**:189-193

[154] Fraternale D, Giamperi L, Ricci D, Rocchi MBL, Guidi L, Epifano F, et al. The effect of triacontanol on micropropagation and on secretory system of *thymus mastichina*. Plant Cell Tissue and Organ Culture. 2003;**74**:87-97

[155] Naeem M, Idrees M, Aftab T, Alam MM, Khan MM, Uddin M, et al. Employing depolymerised sodium alginate, triacontanol and 28-homobrassinolide in enhancing physiological activities, production of essential oil and active components in *Mentha arvensis* L. Industrial Crops and Products. 2014;**55**:272-279

Chapter 5

Revisiting the Medicinal Value of Terpenes and Terpenoids

Deepak Kumar Dash, Chandra Kishore Tyagi, Anil Kumar Sahu and Vaibhav Tripathi

Abstract

Nowadays, plant-based chemicals have drawn the attention of pharmacy researchers due to their potent biological activity against various ailments. In this series, terpenes and terpenoids are gaining popularity among drug researchers gradually. Terpenes are naturally occurring large and varied class of hydrocarbons substances produced by a wide variety of plants including fruits, vegetables, flowers and some animals. Their concentration is generally high in plants. A broad range of the biological properties of terpenoids includes cancer chemo-preventive effects, antimicrobial, antifungal, antiviral, anti-hyperglycemic, anti-inflammatory, anti-parasitic activities and memory enhancers. Terpenoids are usually cyclic unsaturated hydrocarbons, with the altered number of oxygen moieties in the constituent groups attached to the basic isoprene skeleton. Terpenoids are a group of substances that occur in nearly every natural food. Terpenoids display a wide range of biological activities against cancer, malaria, inflammation, tuberculosis and a variety of infectious diseases including viral as well as bacterial. In this chapter, we have emphasized the proven and expected medicinal value of both terpenes and terpenoids.

Keywords: hydrocarbons, chemo-preventive, oxygen moieties, tuberculosis, expected medicinal value

1. Introduction

The role of fruits and vegetables in human nutrition and public health is taken into account in most nutritional recommendations. Fruit and vegetables contain an abundance of various natural compounds that have been associated with the protection and treatment of many ailments.

Terpenes are a huge and diverse category of natural compounds, obtained from a variety of plants, especially conifers, which are characteristic smelling and, in this manner, might have had a defensive function. They are the significant parts of resin and turpentine obtained from the resin. Terpenes are major biosynthetic basic compounds inside every living being. At the point when terpenes are altered chemically, i.e., by oxidation or reframing of the carbon skeleton, the subsequent mixtures are by and large alluded to as terpenoids [1].

IntechOpen

As natural substances, terpenoids are broadly consumed in food, pharmaceuticals, and beauty care products ventures. Indeed, terpenoids are used for the counteraction and treatment of different diseases. Likewise, many investigations have additionally found that terpenoids have numerous expected applications to uncover [2]. This chapter contains updated information about the structure and diverse effects of terpenes and terpenoids.

2. Chemistry and occurrence

Plant biomass is a major potential sustainable source of organic carbon. Terpenes, terpenoids and resin acids are a group of non-polar molecules and share a building block, isoprene or isoprenoid [3], as a common elementary unit (**Figure 1**).

Isoprene, the epitomic terpene substance, is one of the most plentifully ethereal hydrocarbon compounds on Earth attributable to the worldwide plenitude of terpenoid biosynthesis, not the other way around. Around 40% of the biogenic volatile natural compounds transmitted by plants are isoprene, and isoprene is the key hydrocarbon distinguished in human breath. Film crowds breathe out more isoprene when watching scenes of anticipation [5].

Based on ancient scientifically verified data, it is found that the expressions "isoprenoid" and "terpenoid" are applied conversely and that there is still no worldwide accord on terminology.

For instance, a few researchers have named "terpene" to allude just to hydrocarbons dependent on an indispensable number of C5 units, and "terpenoid" or "isoprenoid" to assign the entire class of compound dependent on an intrinsic number of C5 units [6].

Such compounds can be in every way called "terpenes," and the expression "terpenoid" ought to be held for compounds, for example, the steroids, which have differing quantities of carbon molecules, however, are originated from a (C5)n structure [7]. Ruzicka considered this multitude of compounds aggregately to be "terpenic" [8]. Nes and McKean employed the expression "isoprenoid" to portray the entire group [9].

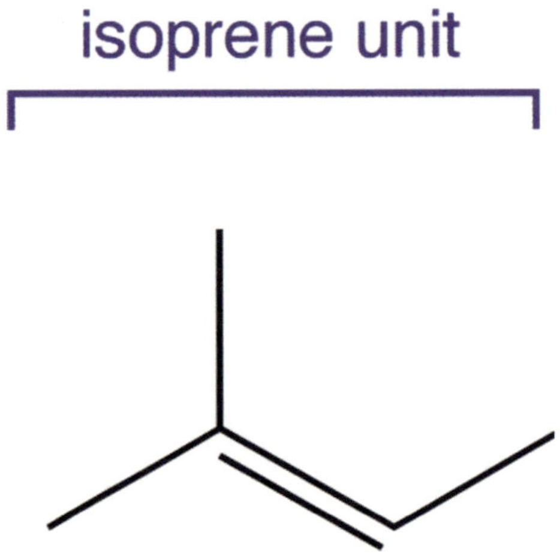

Figure 1.
Structure of isoprene unit [4].

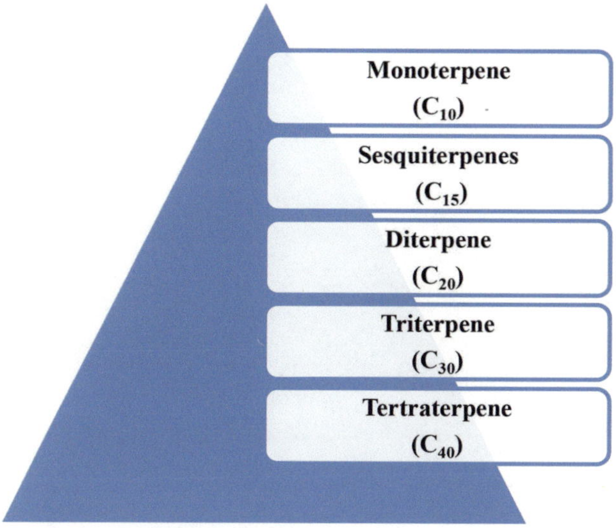

Figure 2.
Classification of terpenes and terpenoids.

Different classes along with carbon units of terpenes and terpenoids are depicted in **Figure 2**.

Terpenes are significant elements of natural oils, which are secondary metabolites synthesized for battling infectious or secreted because of stress conditions. These are extricated from different fragrant plants commonly limited in mild to warm climatic regions like the Mediterranean and tropical nations where they represent a significant piece of the conventional pharmacopeia. They are ethereal, fluid, transparent, and seldom colored, lipid-dissolvable, and dissolvable in natural solvents with a by and large lower thickness than water. They can be isolated from all plant organs, for example, blossoms, buds, leaves, twigs, stems, seeds, roots, wood, fruits or bark, and are accumulated in secretory cells, holes, trenches, glandular trichomes, or epidermic cells [10].

2.1 Monoterpene

The smallest of terpenes are monoterpenes (**Figure 3**). They contain the compound $C_{10}H_{16}$, come from different flowers, fruits and leaves and are known as the main component of essential oils, fragrances and many structural isomers [11]. Monoterpenes are found in natural scents for example α-pinene, which imparts scent to pine trees [12], and limonene from citrus plants [13]. One of the main purposes of monoterpenes is to attract pollinators or to serve the purpose of repelling other organisms from feeding off of plants [14].

2.2 Sesquiterpene

Sesquiterpenes, containing the chemical formula $C_{15}H_{24}$ (**Figure 4**), are much larger compounds than monoterpenes and are much more stable in comparison [15]. Sesquiterpenes are naturally occurring and found in plants, fungi, and insects and act as a defensive mechanism or attract mates with pheromones in insects [16].

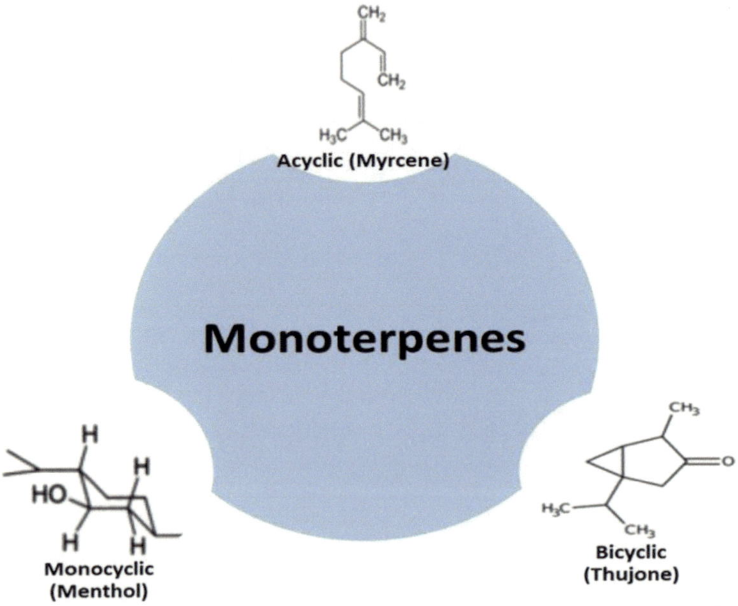

Figure 3.
Sub-classes of monoterpenes with structural example.

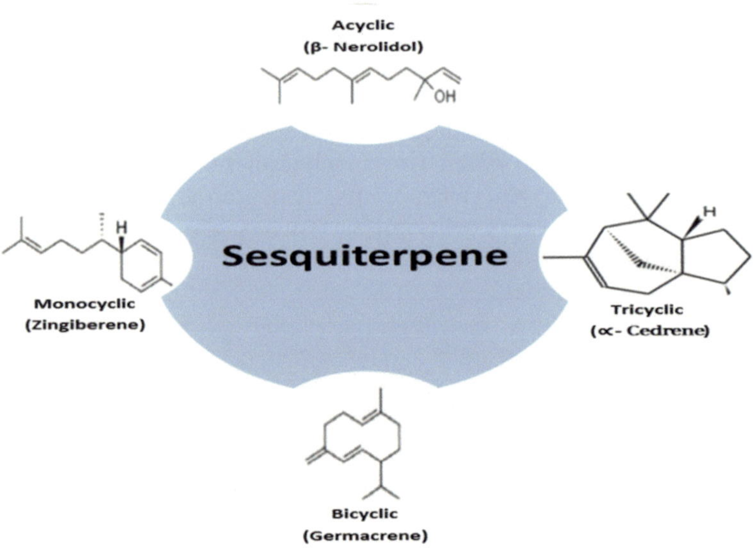

Figure 4.
Classification of sesquiterpenes with structural example.

Gossypol is a sesquiterpene that is present in cotton plants. It has anti-neoplastic properties and might hinder fertility in male people that is the reason it should be taken out from natural oils and different items before human application [17]. Avarol, a sesquiterpenoid that has been displayed to have antifungal and antimicrobial effects, is compelling against AIDS infection [18].

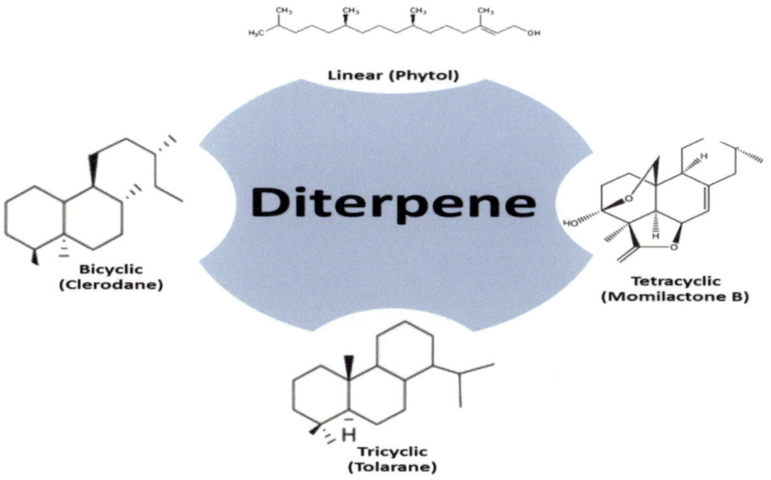

Figure 5.
Sub-classes of diterpenes with structural example.

2.3 Diterpene

Diterpenes are natural substances that contain the atomic skeleton, C20H32 (**Figure 5**) [19]. Diterpenes have physiologically dynamic compounds, for example, plant development chemicals that manage germination, blooming, switch regenerative cycles (from abiogenetic to sexual multiplication) of plants, and vitamin A activity [20]. Cafestol and kahweol are diterpene alcohols that are found in the oil derived from coffee beans [21].

2.4 Triterpene

Triterpenes are composed of three or six isoprene units and have the chemical formula $C_{30}H_{48}$ (**Figure 6**) which includes steroids and sterols with squalene being the biological precursor of all Triterpenes [22]. Triterpenes are produced by animals, plants, and fungi. They play a role as precursors to steroids in animal and plant organisms, and are derived from mevalonic acid [5].

Their properties have been studied for anticancer, antioxidant, antiviral, and anti-atherosclerotic activities [23]. Although, the medicinal uses of tri-terpenes are not quite as recognized as other different types of terpenes but their uses are being continuously investigated by researchers.

2.5 Tetraterpene

Tetraterpenes are also known as carotenoids (**Figure 7**) that have the molecular formula $C_{40}H_{56}$ and can be in the category of terpenes because they are made from isoprene units [24]. They are found in all different types of fungi, bacteria, and plants and are mainly responsible for red, yellow, or orange fat-soluble plant and animal pigments [25]. One of the most crucial and common tetraterpenes is beta-carotene [26].

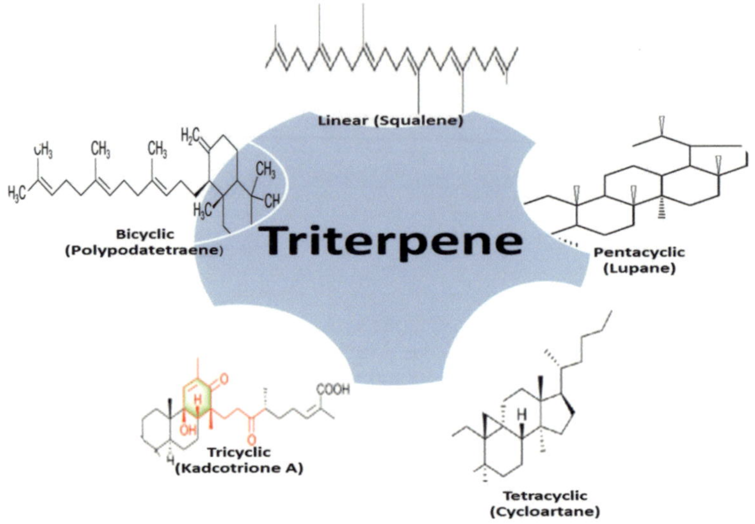

Figure 6.
Classification of Triterpenes.

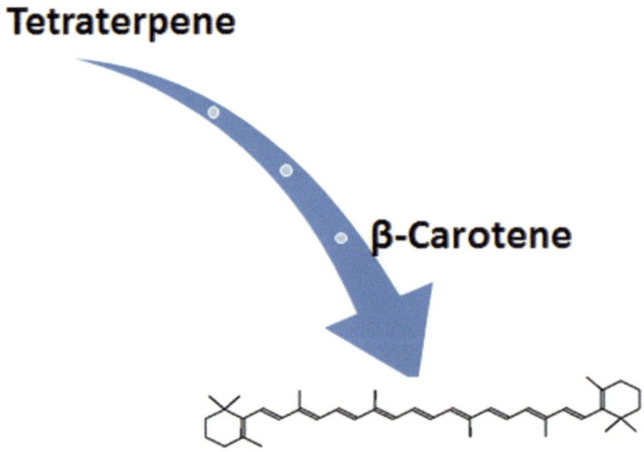

Figure 7.
Structural example of tetraterpene.

3. Plants with terpenoids

Terpenoids, or isoprenoids, are isoprene-based compounds with major jobs in the digestion of all living beings [27]. Varieties of terpenoids are particularly high in plants where many can be viewed as secondary metabolites. Such specific plant terpenoids underlie numerous natural co-operations between plants, creatures, and microorganisms (**Tables 1–4**) [38], going about as allele-synthetics to repulse herbivores, tempt pollinators, or allure herbivore hunters [39]. The development of terpenoids in plants started with the enrollment of genes from primary metabolism and sped up because of the multiplication of cytochrome P450 and terpene synthase gene families in the genomes of plants [40].

Class	Plant source
Monoterpenes	Mentha genus, Cannabis spp.
Sesquiterpenes	Artemisia annua L., Thapsia garganica,
Diterpenes	Taxus brevifolia, Ricinus communis, Euphorbia peplus,
Triterpenes	Azadirachta indica, Khaya grandifolia, Trichilia emetic, Citrus reticulate
Tetraterpenes	Mauritia flexuosa, Brassica oleracea, Crocus sativum L.

Table 1.
List of some plants contains terpenes and terpenoids [28–31].

Class	Insect source
Monoterpene	*Philonthus politus, Zygaena filipendulae, Ips pini*
Sesquiterpene	*Harmonia axyridis, Murgantia histrionica, Myzus persicae*

Table 2.
List of some animals contains terpenes and terpenoids [31–33].

Class	Fungal source
Monoterpenes	*Aspergillus versicolor, Eutypella scoparia, Gelliodes carnosa, Trichoderma asperellum, Thielavia hyalocarpa, Aeromonas hydrophilia, Vibrio anguillarum*
Sesquiterpenes	*Aspergillus fumigatus, Pterocladiella capillacea, Cochliobolus lunatus, Paraconiothyrium sporulosum, Penicillium griseofulvum, Pseudallescheria apiosperma,*
Diterpenes	*Acremonium striatisporum, Aspergillus wentii, Curvularia hawaiiensis, Penicillium commune, Talaromyces purpurogenus, Trichoderma harzianum,*
Triterpenes	*Auxarthron reticulatum,*

Table 3.
List of some fungi contains terpenes and terpenoids [34, 35].

Class	Bacterial source
Sesquiterpenoids	*Streptomyces* strain M491
Diterpenoids	*Streptomyces* strain CNB-982, *Streptomyces sioyaensis, Verrucosispora gifhornensis YM28–088,*
Meroterpenoids	Actinomycete isolates CNH- 099, Erythrobacter sp. strain SNB- 035, Saccharomonos pora sp. CNQ- 490

Table 4.
List of some bacteria contains terpenes and terpenoids [36, 37].

Terpenoid compounds partly mirrors a characteristic history set apart by herbivory stress and other particular tensions forced by creatures, bringing about a wide cluster of functionalized terpenoids in the plant realm pre-chosen for their strong organic activities towards animals [41]. This specific cycle might have been brought about by the overall closeness of protein structures and amino acid sequence among plant and creature proteins, bringing about planting auxiliary metabolites with a natural resemblance for creature proteins by ethicalness of having been delivered by plant bio-catalyst made out of similar amino acids [32].

Terpenoids are dependent on the tetracyclic 6–6–6-5 lanostane carbon skeleton structure a subsection of the terpenome known as the sterolome. The sterolome is assessed to contain about 1000 biogenic derivatives obtained from lanosterol and related particles that do fundamental organic functions across all areas of life on Earth [42].

Many plant terpenoids have been tracked down coincidentally uses in medication and the terpenoids family has been an important wellspring of clinical revelations. However, the testing system is meticulous and asset concentrated. The genuine number of plants terpenoids in nature that might be evaluated for therapeutic applications is obscure however is possibly more than 105, including more than 12,000 from the diterpenoid specifically [43]. While this number is little contrasted with current combinatorial techniques, the lead compound disclosure rate might be altogether higher for plant-based compounds. It is due to a crucial role in chemical, and metabolic processes, many are produced in limited quantities, only in response to a stimulus, or amass solely in particular tissues, requiring microbial multiplication or significant advances by plant rearing and hereditary improvement to get adequate amounts to research clinical benefits [44, 45].

4. Medicinal importance

Terpenoids have an expansive group of clinical activities (**Figure 8**) and are spread everywhere, they have been utilized in conventional medications for ancient times. Numerous compounds can be found commercially, majorly as dietary enhancements; nonetheless, some of them are enrolled as medications.

4.1 Anti- insect activity

Human wellbeing and crop cultivation are mainly affected by insects, and trying to control these bugs the application of chemical bug sprays has become broad. Notwithstanding, this has brought about the improvement of obstruction in these living creatures, human infections, tainting of food, and contamination of the climate. Herbs and medicinal oils like terpenes and terpenoids have been displayed to have a huge potential for bug control like carvacrol, limonene, linalool, 1, 8 cineole, eugenol, and β-ionone; especially against three insects namely lice, cockroaches, and Triatominae bugs [46–48].

4.2 Anti-microbial activity

Antimicrobial properties or the capacity to kill or stop the development of a microorganism in terpenes are normally utilized in customary and current day medication. The accompanying plants produce terpenes that have antimicrobial potential: Pinusponderosa (Pinaceae), flavors (cumin, rosemary, thyme, caraway, clove, and sage), Cretan propolis, Helichrysumitalicum, Rosmarinus officinalis, etc. [49].

There are 52 anti-microbial terpenoids, including hydrocarbons of the oil; aromadendrene (4.4%), limonene (3.8%), α-cedrene (9.6%), β-caryophyllene (4.2%), and α-pinene (10.2%), geranyl acetic acid derivation (4.7%), 2-methylcyclohexyl

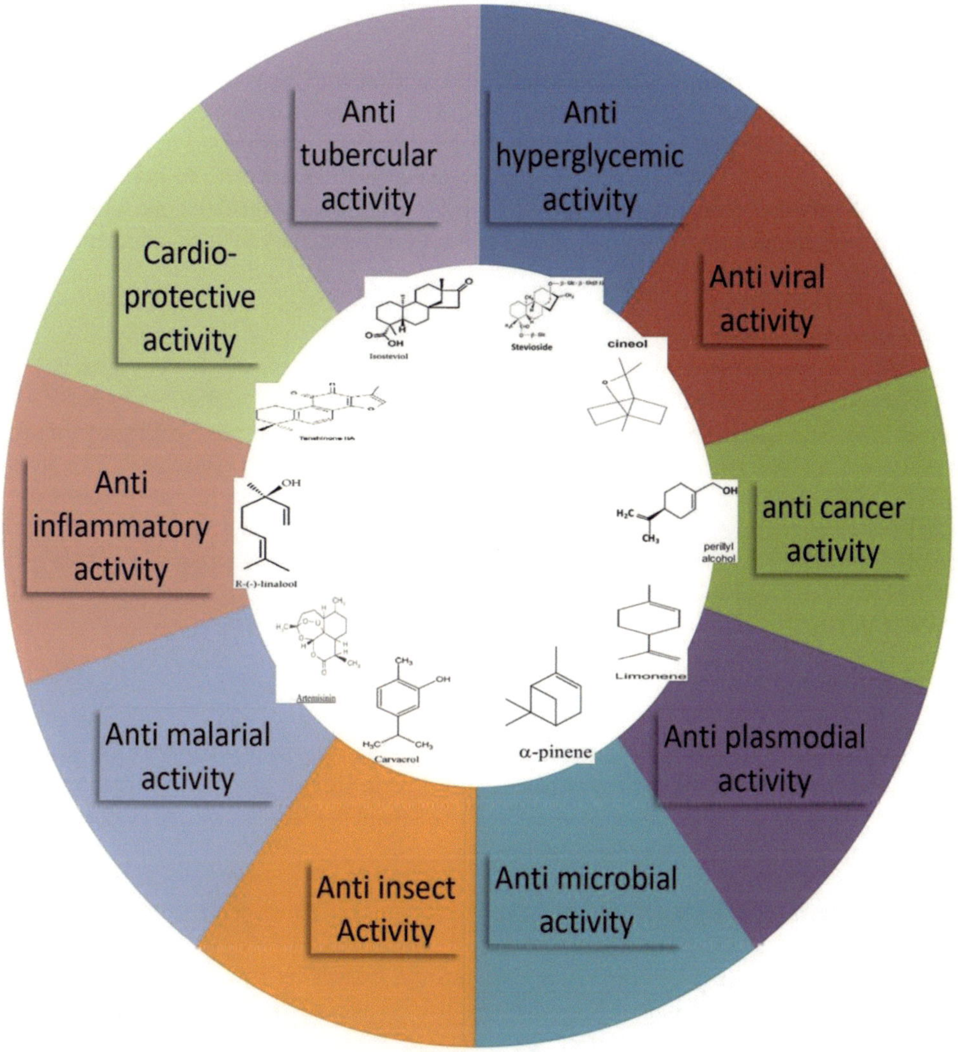

Figure 8.
Reported and traditional therapeutic application of terpenes and terpenoids.

pentanoate (8.3%), 2-methylcyclohexyl octanoate (4.8%), and neryl acetic acid derivation (11.5%) etc. [50, 51].

4.3 Anti- plasmodial activity

Terpenes have been shown to have a favorable anti-plasmodial activity. With the rising malarial infections and drug resistance, terpenes have gained more attention towards it through anti-plasmodial activity. Terpenes have been shown to have a favorable anti-plasmodial activity. With the rising malarial infections and drug resistance, terpenes have gained more attention towards it through anti-plasmodial activity. Different kinds of terpenes show different effects on the parasites. The most common terpenes with anti-plasmodial potential are Beta-myrcene limonene, pinene, caryophyllene, etc. Thus, terpenes could be a safer and a cost-effective alternative for malarial treatment [52, 53].

4.4 Anti-cancer activity

Cancer-related observational studies propose that dietary monoterpenes might be useful in the anticipation and treatment of malignant growths. Among dietary monoterpenes, D-limonene and perillyl alcohol have been displayed to have chemo-preventive and health beneficial properties against numerous human malignant growths. At present they are professed to inhibit fraction-dependent proliferation of skin, lung, mammary, liver, colon, prostate, pancreatic, and stomach carcinomas [54, 55].

4.5 Anti-viral activity

Presently, the antiviral potential of terpenoids is somewhat ineffectively perceived. Consequently, there is a ton of exploration pointed towards finding agents, likewise from natural sources, which could have intense antiviral potential. The new antiviral compounds ought to explicitly restrain the virus and ought not to affect the healthy biological environment of the cell. The quest for natural antiviral moieties has paved the way for the extraction of isoborneol. Potent anti- herpes simplex virus −1 (anti-HSV-1) activities have also been reported for monoterpenes such as cineol and borneol [56, 57].

4.6 Anti-hyperglycemic activity

Type 2 diabetes mellitus is a chronic metabolic disorder that results from reduced first-phase insulin secretion. Stevioside is a diterpene steviol glycoside extracted from leaves of the plant Stevia rebaudiana, which possesses insulinotropic, glucagonostatic, and anti-hyperglycemic effects [58, 59].

4.7 Anti-inflammatory activity

(−)-linalool, a naturally occurring enantiomer, possesses anti-inflammatory activity. Moreover, (−)-linalool and its ester, linalyl acetate, demonstrated analgesic and edema reducing effects [60–62].

4.8 Anti-malarial activity

Artemisinin (sesquiterpene lactone) is secluded from *Artemisia annua* Linn. It is the best antimalarial drug after pyrimethamine, chloroquine, and primaquine, and has the attributes of a high therapeutic index. Afterward, antimalarial medications, for example, artesunate, arteether, and artemether have been isolated by altering the chemistry of artemisinin. Nirolidol likewise has ant-malarial activity [63, 64].

4.9 Cardio-protective activity

Finding a powerful boon for treating the cardiovascular problems is a pressing objective for researchers. Tanshinone IIA (TS) is a functioning moiety separated from the rhizome of Chinese home-grown medication *Salvia miltiorrhiza Bunge*. The most recent discoveries recommend that TS can forestall the emergence of atherosclerosis and the harm and hypertrophy of the heart [65, 66].

4.10 Anti-tubercular activity

Tuberculosis is a very fatal disease to mankind and still the treatment regimen and new drug discovery attract the researchers to reveal a new paradigm in medical science. For the first time, diterpenoid of isosteviol, its binuclear derivatives, tri-terpenoid betulinic, oleanolic, and ursolic acids have been reported to possess anti-tubercular activity, manifested by the molecular docking method. Other natural constituents of the class are Geranylgeraniol, phytanol, escobarine A, escobarine B, furanoditerpenes, salasol A, germacrane, alantolactone, etc. [67, 68].

5. Conclusion

As of now, the clinically evident method of therapeutic activity of numerous terpenoids has not still been clarified. Besides, a relationship of "omics" technology and sub-atomic network pharmacology can be utilized to further affirm the mechanism and structural activity relationship (SAR) of terpenoids. Such study will be a promising step in the development of new medication substances thusly; the compounds with higher interest might be swiftly advanced into new medications, or structurally modified as lead compounds. It is a significant method for the innovative work and development of the medication, and it is additionally a hot spot in the subject of natural product studies. At present, reported terpenoids are about 50,000 inescapable among living beings and major fractions of them are obtained from plants. A few terpenoids are industrially notable for their dietary or therapeutic significance. The new dosage type of terpenoids might be advanced in a blend with the new techniques of pharmaceutics to expand its pharmacological activity. As more terpenoid-based clinical medications will open up, they will assume a vital part in human ailment therapy in forthcoming years. Or terpenoids might be acquainted as additives substances with wellness care items and beauty care products, which has immense market possibilities and money related benefits. We are completely persuaded that it is the beginning stage for the fate of new green science, in light of terpenes and medicinal oils between scientists, industry, and academics.

Author details

Deepak Kumar Dash[1], Chandra Kishore Tyagi[2], Anil Kumar Sahu[1]
and Vaibhav Tripathi[1*]

1 Royal College of Pharmacy, Raipur, India

2 College of Pharmacy, Sri Satya Sai University of Technology and Medical Sciences, Sehore, India

*Address all correspondence to: vaibhu.07@gmail.com

References

[1] Yang W, Chen X, Li Y, Guo S, Wang Z, Yu X. Advances in pharmacological activities of terpenoids. Natural Product Communications. 2020;**15**(3): 1934578X20903555

[2] Sahu P, Bhowmick AK, Kali G. Terpene based elastomers: Synthesis, properties, and applications. PRO. 2020;**8**(5):553

[3] Karunanithi PS, Zerbe P. Terpene synthases as metabolic gatekeepers in the evolution of plant terpenoid chemical diversity. Frontiers in Plant Science. 2019;**10**:1166

[4] Thomsett MR, Storr TE, Monaghan OR, Stockman RA, Howdle SM. Progress in the synthesis of sustainable polymers from terpenes and terpenoids. Green Materials. 2016;**4**(3):115-134

[5] Hillier SG, Lathe R. Terpenes, hormones and life: Isoprene rule revisited. Journal of Endocrinology. 2019;**242**(2):R9-R22

[6] Devarenne TP. Terpenoids: Higher. In: Encyclopedia of Life Sciences (ELS). Chichester: John Wiley & Sons, Ltd; 2009

[7] Petrović J, Stojković D, Soković M. Terpene core in selected aromatic and edible plants: Natural health improving agents. Advances in Food and Nutrition Research. 2019;**90**:423-451

[8] Medbouhi A, Benbelaïd F, Djabou N, Beaufay C, Bendahou M, Quetin-Leclercq J, et al. Essential oil of algerian Eryngium campestre: Chemical variability and evaluation of biological activities. Molecules. 2019;**24**(14):2575

[9] Nes WR, McKean ML. Biochemistry of Steroids and Other Isopentenoids. Baltimore, United States: University Park Press; 1977. pp. 690

[10] Touaibia M, Boutekedjiret C, Perino S, Chemat F. Natural terpenes as building blocks for green chemistry. In: Plant Based "Green Chemistry 2.0". Singapore: Springer; 2019. pp. 171-195

[11] Dehsheikh AB, Sourestani MM, Dehsheikh PB, Mottaghipisheh J, Vitalini S, Iriti M. Monoterpenes: Essential oil components with valuable features. Mini reviews in medicinal chemistry. 2020;**20**(11):958-974

[12] Cox-Georgian D, Ramadoss N, Dona C, Basu C. Therapeutic and medicinal uses of terpenes. Medicinal Plants. 2020;**11**:333-359

[13] Thielmann J, Muranyi P. Review on the chemical composition of Litsea cubeba essential oils and the bioactivity of its major constituents citral and limonene. Journal of Essential Oil Research. 2019;**31**(5):361-378

[14] Tetali SD. Terpenes and isoprenoids: a wealth of compounds for global use. Planta. 2019;**249**(1):1-8

[15] Wang Y, Ma Y, Li X, Kuang BY, Huang C, Tong R, et al. Monoterpene and Sesquiterpene α-Hydroxy Organosulfates: Synthesis, MS/MS Characteristics, and Ambient Presence. Environmental Science & Technology. 2019;**53**(21):12278-12290

[16] Sharma A, Sandhi RK, Reddy GV. A review of interactions between insect biological control agents and semiochemicals. Insects. 2019;**10**(12):439

[17] Cao H, Sethumadhavan K, Bland JM. Isolation of cottonseed extracts that affect human cancer cell growth. Scientific Reports. 2018;**8**(1):1-2

[18] Ghosh S, Roy K, Pal C. Terpenoids against infectious diseases. In: Terpenoids against human diseases. Vol. 11. Boca Raton: CRC Press; 2019. pp. 187-208

[19] Sharma M, DwivediI J, Kumar B, Singh B, Rawat A. Plant-Based Secondary Metabolites for Health Benefits: Classification and Processing. In: Phytochemicals from Medicinal Plants: Scope, Applications, and Potential Health Claims. Vol. 33. Burlington, Canada: Apple Academic Press Inc.; 2019

[20] Lange BM. The evolution of plant secretory structures and emergence of terpenoid chemical diversity. Annual Review of Plant Biology. 2015;**66**:139-159

[21] Novaes FJ, Bayan FC, Neto FR, Resende CM. The occurrence of cafestol and kahweol diterpenes in different coffee brews. Coffee Science. 2019;**14**(2):265-280

[22] Ávila EC. Terpenoids in marine heterobranch molluscs. Marine Drugs. 2020;**18**(3):162-200

[23] Bharadwaj S, Lee KE, Dwivedi VD, Yadava U, Panwar A, Lucas SJ, et al. Discovery of Ganoderma lucidum triterpenoids as potential inhibitors against Dengue virus NS2B-NS3 protease. Scientific Reports. 2019;**9**(1):1-2

[24] Mohiuddin AK. Chemistry of Secondary Metabolites. Ann Clin Toxicol. 2019;**2**(1):1014

[25] Jurić S, Jurić M, Król-Kilińska Ż, Vlahoviček-Kahlina K, Vinceković M, Dragović-Uzelac V, et al. Sources, stability, encapsulation and application of natural pigments in foods. Food Reviews International. 2020;**29**:1-56

[26] Rammuni MN, Ariyadasa TU, Nimarshana PH, Attalage RA. Comparative assessment on the extraction of carotenoids from microalgal sources: Astaxanthin from H. pluvialis and β-carotene from D. salina. Food Chemistry. 2019;**277**:128-134

[27] Ghorbanpour M, Hadian J, Nikabadi S, Varma A. Importance of medicinal and aromatic plants in human life. Medicinal Plants and Environmental Challenges. 2017;**31**:1

[28] Kishkentayeva AS. Production technologies of pharmacologically active sesquiterpene lactones. Eurasian Chemico-Technological Journal. 2018;**20**(4):325-333

[29] Warra AA. Castor (Ricinus communis L.) plant: Medicinal, environmental and industrial applications. International Journal of Agricultural and Environmental Sciences. 2018;**3**(5):78

[30] Goyal L, Kaushal S. Evaluation of chemical composition and antioxidant potential of essential oil from citrus reticulata fruit peels. Advances in Research. 2018;**31**:1-9

[31] Bataglion GA, Paz WH, Adrião AA, da Rocha Albuquerque JM, da Silva FM, Numa IA, et al. Bioactive compounds of buriti fruit (Mauritia flexuosa Lf). Bioactive Compounds in Underutilized Fruits and Nuts. 2020:411-436

[32] Beran F, Köllner TG, Gershenzon J, Tholl D. Chemical convergence between plants and insects: Biosynthetic origins and functions of common secondary metabolites. New Phytologist. 2019;**223**(1):52-67

[33] Miura T, Maekawa K. The making of the defensive caste: Physiology, development, and evolution of the soldier differentiation in termites. Evolution & Development. 2020;**22**(6): 425-437

[34] Jiang M, Wu Z, Guo H, Liu L, Chen S. A review of terpenes from marine-derived fungi: 2015-2019. Marine drugs. 2020;**18**(6):321

[35] Sorvari J, Hartikainen S. Terpenes and fungal biomass in the nest mounds of Formica aquilonia wood ants. European Journal of Soil Biology. 2021;**105**:103336

[36] Gozari M, Alborz M, El-Seedi HR, Jassbi AR. Chemistry, biosynthesis and biological activity of terpenoids and meroterpenoids in bacteria and fungi isolated from different marine habitats. European Journal of Medicinal Chemistry. 2021;**210**:112957

[37] Reddy GK, Leferink NG, Umemura M, Ahmed ST, Breitling R, Scrutton NS, et al. Exploring novel bacterial terpene synthases. PLoS One. 2020;**15**(4):e0232220

[38] Bergman ME, Davis B, Phillips MA. Medically useful plant terpenoids: Biosynthesis, occurrence, and mechanism of action. Molecules. 2019;**24**(21):3961

[39] Takabayashi J, Shiojiri K. Multifunctionality of herbivory-induced plant volatiles in chemical communication in tritrophic interactions. Current Opinion in Insect Science. 2019;**32**:110-117

[40] El-Sayed AS, El-Sayed MT, Rady A, Zein N, Enan G, Shindia A, et al. Exploiting the biosynthetic potency of taxol from fungal endophytes of conifers plants. Genome Mining and Metabolic Manipulation. Molecules. 2020;**25**(13):3000

[41] Schürrle K. History, current state, and emerging applications of industrial biotechnology. Advances in Biochemical Engineering/biotechnology. 2019;**123**:13-51

[42] Huang PW, Wang LR, Geng SS, Ye C, Sun XM, Huang H. Strategies for enhancing terpenoids accumulation in microalgae. Applied Microbiology and Biotechnology. 2021;**14**:1-2

[43] Inabuy FS. Biosynthesis of Plant Specialized Diterpenoids: Leveraging Metabolic and Transcriptomic Profiling for Gene Discovery. United States; Published by ProQuest LLC, Washington State University; 2017

[44] Nzila A. Biodegradation of high-molecular-weight polycyclic aromatic hydrocarbons under anaerobic conditions: Overview of studies, proposed pathways and future perspectives. Environmental Pollution. 2018;**239**:788-802

[45] Darvishpor S, Hosseini A, Davoodi A, Salehifar E, Akbari J, Azadbakht M. A review on medicinal plants used for nausea and vomiting in Persian Medicine. Global Journal of Medical Research. 2018;**18**(1):29-45

[46] Tak JH, Isman MB. Penetration-enhancement underlies synergy of plant essential oil terpenoids as insecticides in the cabbage looper. Trichoplusia ni. Scientific reports. 2017;7(1):1-1

[47] Sarma R, Adhikari K, Mahanta S, Khanikor B. Combinations of plant essential oil based terpene compounds as larvicidal and adulticidal agent against Aedes aegypti (Diptera: Culicidae). Scientific Reports. 2019;**9**(1):1-2

[48] Tripathi AK, Upadhyay S, Bhuiyan M, Bhattacharya PR. A review on prospects of essential oils as

biopesticide in insect-pest management. Journal of Pharmacognosy and phytotherapy. 2009;**1**(5):052-063

[49] Akthar MS, Degaga B, Azam T. Antimicrobial activity of essential oils extracted from medicinal plants against the pathogenic microorganisms: A review. Journal Issues ISSN. 2014; **2350**:1588

[50] Mahizan NA, Yang SK, Moo CL, Song AA, Chong CM, Chong CW, et al. Terpene derivatives as a potential agent against antimicrobial resistance (AMR) pathogens. Molecules. 2019;**24**(14):2631

[51] Gutiérrez-del-Río I, Fernández J, Lombó F. Plant nutraceuticals as antimicrobial agents in food preservation: Terpenoids, polyphenols and thiols. International journal of antimicrobial agents. 2018;**52**(3):309-315

[52] Singh E, Khan RJ, Jha RK, Amera GM, Jain M, Singh RP, et al. A comprehensive review on promising anti-viral therapeutic candidates identified against main protease from SARS-CoV-2 through various computational methods. Journal of Genetic Engineering and Biotechnology. 2020;**18**(1):1-2

[53] Shoker RM. A review article: The importance of the major groups of plants secondary metabolism phenols, alkaloids, and terpenes. International Journal for Research in Applied Sciences and Biotechnology. 2020;**7**(5):354-358

[54] Kuttan G, Pratheeshkumar P, Manu KA, Kuttan R. Inhibition of tumor progression by naturally occurring terpenoids. Pharmaceutical Biology. 2011;**49**(10):995-1007

[55] Sharma SH, Thulasingam S, Nagarajan S. Terpenoids as anti-colon cancer agents–A comprehensive review on its mechanistic perspectives. European journal of pharmacology. 2017;**795**:169-178

[56] Rezanka T, Siristova L, Sigler K. Sterols and triterpenoids with antiviral activity. Anti-Infective Agents in Medicinal Chemistry (Formerly Current Medicinal Chemistry-Anti-Infective Agents). 2009;**8**(3):193-210

[57] Flechas MC, Ocazionez RE, Stashenko EE. Evaluation of in vitro antiviral activity of essential oil compounds against dengue virus. Pharmacognosy Journal. 2018;**10**(1):55-59

[58] Panigrahy SK, Bhatt R, Kumar A. Targeting type II diabetes with plant terpenes: the new and promising antidiabetic therapeutics. Biologia. 2021;**76**(1):241-254

[59] Kaur KK, Allahbadia G, Singh M. Monoterpenes-a class of terpenoid group of natural products as a source of natural antidiabetic agents in the future-A Review. CPQ Nutrition. 2019;**3**(4):1-21

[60] Prakash VE. Terpenoids as source of anti-inflammatory compounds. Asian Journal of Pharmaceutical and Clinical Research. 2017;**10**(3):68-76

[61] Vega RJ, Xolalpa NC, Castro AJ, Pérez González C, Pérez Ramos J, Pérez GS. Terpenes from natural products with potential anti-inflammatory activity. Terpenes and Terpenoids. IntechOpen. 2018;**5**:59-85

[62] Kim T, Song B, Cho KS, Lee IS. Therapeutic potential of volatile terpenes and terpenoids from forests for inflammatory diseases. International Journal of Molecular Sciences. 2020;**21**(6):2187

[63] Saito AY, Rodriguez AA, Vega DS, Sussmann RA, Kimura EA, Katzin AM. Antimalarial activity of the terpene

nerolidol. International journal of antimicrobial agents. 2016;**48**(6):641-646

[64] Silva GN, Rezende LC, Emery FS, Gosmann G, Gnoatto SC. Natural and semi synthetic antimalarial compounds: Emphasis on the terpene class. Mini reviews in Medicinal Chemistry. 2015;**15**(10):809-836

[65] Mahomoodally F, Ramjuttun P. Terpenoids Against Cardiovascular Diseases. In: Terpenoids Against Human Diseases. London, New York: CRC Press; 2019. pp. 209-231

[66] Silva EA, Carvalho JS, Guimarães AG, Barreto RD, Santos MR, Barreto AS, et al. The use of terpenes and derivatives as a new perspective for cardiovascular disease treatment: A patent review (2008-2018). Expert Opinion on Therapeutic Patents. 2019;**29**(1):43-53

[67] Kataev VE, Khaybullin RN, Garifullin BF, Sharipova RR. New targets for growth inhibition of mycobacterium tuberculosis: Why do natural terpenoids exhibit antitubercular activity? Russian Journal of Bioorganic Chemistry. 2018;**44**(4):438-452

[68] Sansinenea E, Ortiz A. Antitubercular natural terpenoids: Recent developments and syntheses. Current Organic Synthesis. 2014;**11**(4): 545-591